Applied Mathematical Sciences

EDITORIAL STATEMENT

The mathematization of all sciences, the fading of traditional scientific boundaries, the impact of computer technology, the growing importance of mathematical-computer modelling and the necessity of scientific planning all create the need both in education and research for books that are introductory to and abreast of these developments.

The purpose of this series is to provide such books, suitable for the user of mathematics, the mathematician interested in applications, and the student scientist. In particular, this series will provide an outlet for material less formally presented and more anticipatory of needs than finished texts or monographs, yet of immediate interest because of the novelty of its treatment of an application or of mathematics being applied or lying close to applications.

The aim of the series is, through rapid publication in an attractive but inexpensive format, to make material of current interest widely accessible. This implies the absence of excessive generality and abstraction, and unrealistic idealization, but with quality of exposition as a goal.

Many of the books will originate out of and will stimulate the development of new undergraduate and graduate courses in the applications of mathematics. Some of the books will present introductions to new areas of research, new applications and act as signposts for new directions in the mathematical sciences. This series will often serve as an intermediate stage of the publication of material which, through exposure here, will be further developed and refined. These will appear in conventional format and in hard cover.

MANUSCRIPTS

The Editors welcome all inquiries regarding the submission of manuscripts for the series. Final preparation of all manuscripts will take place in the editorial offices of the series in the Division of Applied Mathematics, Brown University, Providence, Rhode Island.

SPRINGER-VERLAG NEW YORK INC., 175 Fifth Avenue, New York, N. Y. 10010

Applied Mathematical Sciences | Volume 63

Applied Mathematical Sciences

1. John: **Partial Differential Equations,** 4th ed.
2. Sirovich: **Techniques of Asymptotic Analysis.**
3. Hale: **Theory of Functional Differential Equations,** 2nd ed.
4. Percus: **Combinatorial Methods.**
5. von Mises/Friedrichs: **Fluid Dynamics.**
6. Freiberger/Grenander: **A Short Course in Computational Probability and Statistics.**
7. Pipkin: **Lectures on Viscoelasticity Theory.**
9. Friedrichs: **Spectral Theory of Operators in Hilbert Space.**
11. Wolovich: **Linear Multivariable Systems.**
12. Berkovitz: **Optimal Control Theory.**
13. Bluman/Cole: **Similarity Methods for Differential Equations.**
14. Yoshizawa: **Stability Theory and the Existence of Periodic Solutions and Almost Periodic Solutions.**
15. Braun: **Differential Equations and Their Applications,** 3rd ed.
16. Lefschetz: **Applications of Algebraic Topology.**
17. Collatz/Wetterling: **Optimization Problems.**
18. Grenander: **Pattern Synthesis: Lectures in Pattern Theory, Vol I.**
20. Driver: **Ordinary and Delay Differential Equations.**
21. Courant/Friedrichs: **Supersonic Flow and Shock Waves.**
22. Rouche/Habets/Laloy: **Stability Theory by Liapunov's Direct Method.**
23. Lamperti: **Stochastic Processes: A Survey of the Mathematical Theory.**
24. Grenander: **Pattern Analysis: Lectures in Pattern Theory, Vol. II.**
25. Davies: **Integral Transforms and Their Applications,** 2nd ed.
26. Kushner/Clark: **Stochastic Approximation Methods for Constrained and Unconstrained Systems.**
27. de Boor: **A Practical Guide to Splines.**
28. Keilson: **Markov Chain Models—Rarity and Exponentiality.**
29. de Veubeke: **A Course in Elasticity.**
30. Sniatycki: **Geometric Quantization and Quantum Mechanics.**
31. Reid: **Sturmian Theory for Ordinary Differential Equations.**
32. Meis/Markowitz: **Numerical Solution of Partial Differential Equations.**
33. Grenander: **Regular Structures: Lectures in Pattern Theory, Vol. III.**
34. Kevorkian/Cole: **Perturbation Methods in Applied Mathematics.**
35. Carr: **Applications of Centre Manifold Theory.**
36. Bengtsson/Ghil/Källén: **Dynamic Meterology: Data Assimilation Methods.**
37. Saperstone: **Semidynamical Systems in Infinite Dimensional Spaces.**
38. Lichtenberg/Lieberman: **Regular and Stochastic Motion.**

(continued on inside back cover)

Johan Grasman

Asymptotic Methods for Relaxation Oscillations and Applications

With 85 Illustrations

Springer-Verlag
New York Berlin Heidelberg
London Paris Tokyo

Johan Grasman
Department of Mathematics
State University of Utrecht
3508 TA Utrecht
The Netherlands

AMS Subject Classification: 34C15, 34EXX, 58F13, 60J70, 92A09

Library of Congress Cataloging in Publication Data
Grasman, Johan.
Asymptotic methods for relaxation
oscillations and applications.
(Applied mathematical sciences ; v. 63)
Bibliography: p.
Includes indexes.
1. Asymptotic expansions. 2. Differential
equations, Nonlinear—Asymptotic theory.
3. Oscillations. I. Title. II. Title:
Relaxation oscillations and applications.
III. Series: Applied mathematical sciences
(Springer-Verlag New York Inc.)
QA1.A647 vol. 63 510 s 87-4556
[QA372] [515.3′55]

Printed and bound by Arcata Graphics/Halliday, West Hanover, Massachusetts.
Printed in the United States of America

9 8 7 6 5 4 3 2 1

ISBN 0-387-96513-0 Springer-Verlag New York Berlin Heidelberg
ISBN 3-540-96513-0 Springer-Verlag Berlin Heidelberg New York

Dedicated to Ank, Laurens and Stefan

Preface

In various fields of science, notably in physics and biology, one is confronted with periodic phenomena having a remarkable temporal structure: it is as if certain systems are periodically reset in an initial state. A paper of Van der Pol in the Philosophical Magazine of 1926 started up the investigation of this highly nonlinear type of oscillation for which Van der Pol coined the name "relaxation oscillation".

The study of relaxation oscillations requires a mathematical analysis which differs strongly from the well-known theory of almost linear oscillations. In this monograph the method of matched asymptotic expansions is employed to approximate the periodic orbit of a relaxation oscillator. As an introduction, in chapter 2 the asymptotic analysis of Van der Pol's equation is carried out in all detail. The problem exhibits all features characteristic for a relaxation oscillation. From this case study one may learn how to handle other or more generally formulated relaxation oscillations. In the survey special attention is given to biological and chemical relaxation oscillators.

In chapter 2 a general definition of a relaxation oscillation is formulated. Essential is the existence of a phase of rapid change which can be related to the presence of a small parameter in the equation. An investigation of chaotic and stochastic oscillations completes the analysis of free oscillations of chapter 2. In chapter 3 the dynamics of coupled oscillators is analyzed. The coupling leads to entrainment phenomena for which one may find many applications in biology. The existence of phase waves in a system of spatially distributed coupled oscillators is an example of such an application. In chapter 4 subharmonic and chaotic solutions of the Van der Pol equation with a sinusoidal forcing term are constructed. This problem, formulated by Littlewood, intrigued mathematicians over the last forty years. The horse-shoe mapping of Smale was intended to be used in the analysis of this problem. Later on it was noticed that chaos is present at a much wider scale.

My approach to the analysis of relaxation oscillations is one of an applied mathematician. New developments in the theory of nonlinear dynamical systems made it possible to describe qualitatively phenomena such as chaotic dynamics of physical and biological systems. The question arose whether in such cases quantitative approximations can be made. Using powerful tools from the theory of singular perturbations we came into the position to construct matched asymptotic approximations and to relate them to the results of the qualitative theories. For biologists and physists it is worthy to get acquainted with the outcome of these investigations: entrainment phenomena in systems of coupled biological oscillators can be quantified and chaotic dynamics of driven oscillators can be computed!

I have attempted to give a survey of the literature following the historical developments in the field and to make the bibliography as complete as possible. The book gives an overview of the work that has been done in this field over the last sixty years. There may be ommisions and the contents is perhaps somewhat out of in balance when I deal with the contributions of our group in the Netherlands. One last remark has to be made about the historical element. While reading the papers of Van der Pol, I got impressed by his intuitive judgement of the importance of certain physical phenomena, his ability to analyse them mathematically and by his charming and effective ways of conveying his observations to an audience of scientists and medical investigators.

This study of relaxation oscillations will aquaint the reader with the modeling of periodic phenomena in physics and biology. It, moreover, demonstrates how the modern theory of dynamical systems is applied to a particular type of nonlinear oscillation. In order to explore the distinction between chaos and noise the effect of stochastic perturbations upon the oscillator and its period will be analyzed. Obviously, this broad approach with a diversity of mathematical techniques makes it difficult to study in depth all theories that are brought up. The reader should consult the literature if he wishes to acquire a better background knowledge of a mathematical topic such as the theory of stochastic differential equations or the use of symbolic dynamics.

This book can be read by one who has some basic knowledge of differential equations and asymptotic methods. In the appendices I reviewed concepts of these theories, which are useful to have at hand. The presentation of this material is too brief to use it as an introduction in this field. As a compensation appropriate references are given.

Biologists may, in particular, be interested in chapters 1,2 and 3 with exclusion of sections 2.1.3-2.1.5, 2.2.2-2.2.4 and 3.2.1-3.2.3. Indeed, the book can be read in this way.

I am indebted to Wiktor Eckhaus Michael Ghil, Hans Lauwerier, Jos Roerdinck, Nico Temme and Ferdinand Verhulst for their useful suggestions which have helped to improve the content and clarity of the manuscript. Moreover, I am grateful to the Centre of Mathematics and Computer Science, that gave me the opportunity to carry out the mathematical investigations which has led to the present monograph. The Centre also provided me with all facilities for computing and for reproduction of text and figures. I wish to mention the assistance of Boudewijn de Kerf in the use of computer graphics. Miss Mini Middelberg and miss Sandra Dorrestijn took care of the typing of the manuscript; they did an excellent job. The figures with perspective has been drawn by Tobias Baanders; his fine style will certainly help the reader to perceive the course of trajectories in phase space.

Amsterdam, August 1986 Johan Grasman

Contents

Appendices

1. INTRODUCTION

Intuitively the dynamics of a relaxation oscillation is easily understood from a simple mechanical system as the see-saw of fig. 1.0.1 with a water reservoir at one side. As the amount of water exceeds the weight at the other side, the see-saw flips. Then the reservoir is emptied and returns to its original position. In the applied sciences relaxation oscillations are most frequently met in biochemical and biological systems. A similar phenomenon is observed for these systems: during a short time interval of the cycle one or more components of the biological system may exhibit a fast change in their density.

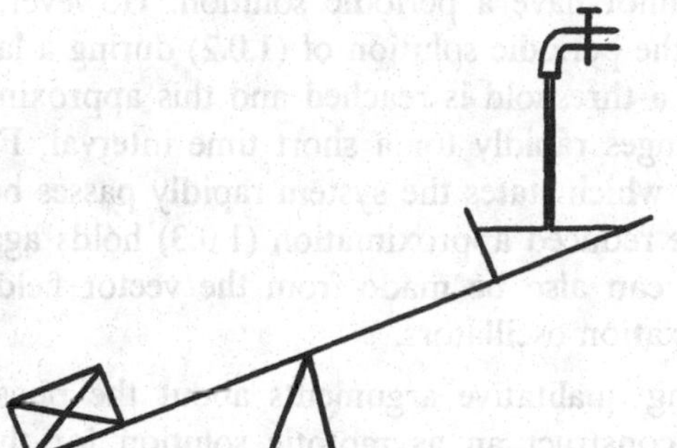

Fig.1.0.1 A typical relaxation oscillator: the see-saw with a water reservoir at one side. The period depends on the rate of inflow and the volume of water in the state of balance.

For almost linear oscillators a more or less generally accepted mathematical theory exists, see Bogoliubov and Mitropolsky (1961). A one degree of freedom system takes in that case the form

$$\frac{d^2x}{dt^2}+x = \epsilon f\left(x,\frac{dx}{dt}\right), \quad 0<\epsilon<<1. \qquad (1.0.1)$$

When ϵ is not small this theory cannot be applied. In general, no asymptotic approximations can be constructed and solutions have to be found numerically. However, there is an exception: for systems of the type

$$\epsilon \frac{d^2x}{dt^2} = f\,(x, \frac{dx}{dt}), \quad 0<\epsilon<<1 \tag{1.0.2}$$

again asymptotic approximations can be made. Letting $\epsilon \rightarrow 0$ we obtain the equation

$$f(x, \frac{dx}{dt}) = 0, \tag{1.0.3}$$

which by itself cannot have a periodic solution. However, it may be a good approximation of the periodic solution of (1.0.2) during a large time interval of the cycle. Then as a threshold is reached and this approximation breaks down the variable x changes rapidly for a short time interval. From the see-saw of fig. 1.0.1 it is clear which states the system rapidly passes before it returns in a state for which the reduced approximation (1.0.3) holds again. As we will see such a conclusion can also be made from the vector field in state space for other types of relaxation oscillators.

Instead of using qualitative arguments about the phase of rapid change, one may as well construct an asymptotic solution for this time interval by stretching the time variable. This solution must match the one for the phase in which approximation (1.0.3) holds. This technique is known as the method of matched asymptotic expansions. It originates from fluid mechanics, where it is used to analyse boundary layer phenomena. In the last ten years a general theory for this class of problems evolved, see Kevorkian and Cole (1981), O'Malley (1982) and Eckhaus (1979). Characteristic for problems in singular perturbation theory is the small parameter that multiplies the highest derivative, see formula (1.0.2).

In the following sections of this introduction we give an overview of the occurrence of relaxation oscillations in the applied science. We distinguish between free, forced and coupled oscillations, which comes up as a quite natural classification. In the following chapters these oscillations are analyzed mathematically and reference is made to the applications mentioned in this introduction. Moreover, some examples are worked out.

1.1 The Van der Pol oscillator

Relaxation oscillations were observed the first time by Van der Pol (1926), who studied properties of a triode circuit. Such a system exhibits self-sustained oscillations with an amplitude independent of the starting conditions. For certain values of the system parameters the oscillation is almost sinusoidal, while in a different range the oscillation shows abrupt changes, see fig. 1.1.1. In the last case the period turns out to be about proportional to a large parameter of the system. The name "relaxation oscillation" refers to this characteristic time constant of the system. Le Corbeiller (1931) took over this name. In fig. 1.1.2 the triode circuit of Van der Pol is given. The system satisfies the following equation

$$L\frac{dI}{dt} + RI + V = M\frac{dI_a}{dt}, \tag{1.1.1}$$

where L is the selfinductance, M the mutual inductance, R a resistance and I and V, respectively, a current and the grid voltage. Assuming that the grid current is negligible we have $I = CV'(t)$, where C is a capacitance. The triode characteristic is idealized as $I_a = V - 1/3V^3$. Then by substituting

$$V = x\sqrt{1 - RC/M}, \quad t = \tau\sqrt{LC} \tag{1.1.2ab}$$

we obtain the well-known Van der Pol equation

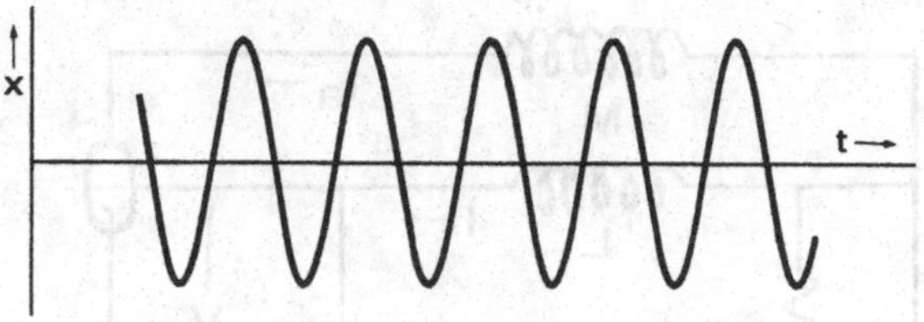

(a) $\nu = 0.1$

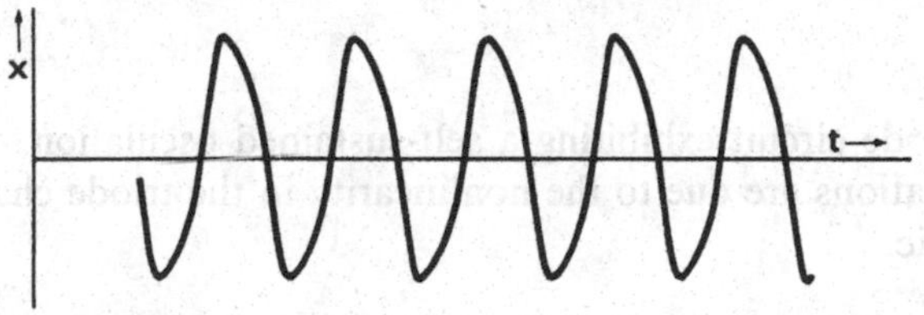

(b) $\nu = 1.0$

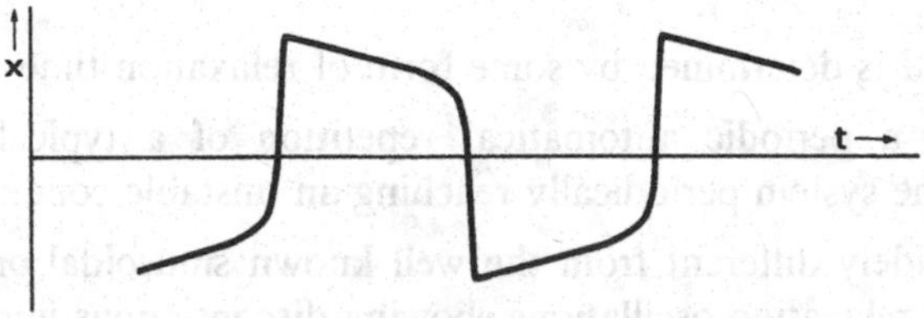

(c) $\nu = 10.$

Fig.1.1.1 Periodic solution of the Van der Pol equation for different values of the parameter. For $\nu=0.1$ the oscillation is almost sinusoidal with a period of about 2π, while for $\nu=10.$ the oscillation is almost discontinuous with a period of about $(3-2\ln 2)\nu$.

$$\frac{d^2x}{d\tau^2} + \nu(x^2-1)\frac{dx}{d\tau} + x = 0 \tag{1.1.3}$$

with

$$\nu = M/\sqrt{LC} - R\sqrt{C/L}. \tag{1.1.4}$$

Van der Pol (1926) also investigated Eq. (1.1.3) asymptotically for $0<\nu<<1$ and gave numerical solutions for the phase-plane as well as the physical plane for three characteristic choices of ν. For $\nu>>1$ the periodic solution of (1.1.3) behaves as a relaxation oscillation.

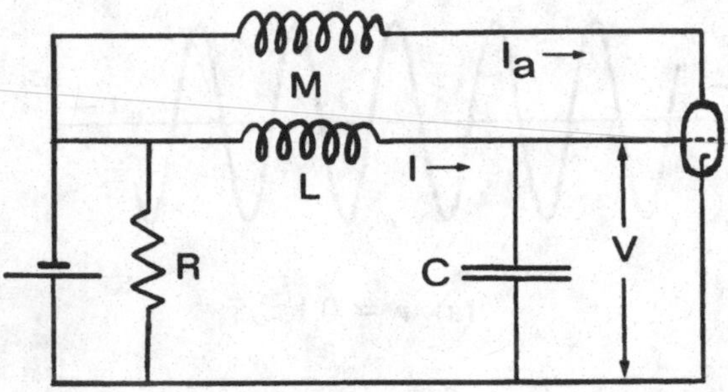

Fig.1.1.2 A triode circuit exhibiting a self-sustained oscillation. The oscillations are due to the nonlinearity in the triode characteristic.

Later on Van der Pol (1940) listed the following properties of relaxation oscillations:

a. their time period is determined by some form of relaxation time
b. they constitute a periodic automatical repetition of a typical aperiodic phenomenon, the system periodically reaching an unstable condition
c. their form is widely different from the well known sinusoidal or harmonic oscillations, the relaxation oscillations showing discontinuous jumps
d. their time period is not far as constant as in the case of sinusoidal oscillations because external circumstances such as temperature variation may much easier influence a relaxation time than a sinusoidal time period
e. they were found from a nonlinear differential equation implying the presence of a threshold value resulting in the applicability of the all-or-nothing law.

In the same paper Van der Pol quotes the English physiologist A.V. Hill (1933), who, in a study on nerve activity, states that relaxation oscillations are the type of oscillations that gouverns all periodic phenomena in physiology. Besides this field of application and that of electronic networks Van der Pol expects that relaxation oscillations are also present in population cycles and refers to Volterra and Lotka for further details. This passage in the paper of Van der Pol was noticed after the content of this book was more or less fixed. It is surprising to see how well the thoughts of Van der Pol are reflected in this book.

1.2 Mechanical prototypes of relaxation oscillators

The self-sustained oscillation in the triode circuit of the preceding section is due to the nonlinearity in the I, V-characteristic of the triode. It is not so difficult to construct mechanical devices in which the effect of a nonlinearity is better visible in the action of the system. In fig. 1.2.1a the see-saw of fig.1.0.1 is depicted again. The dynamics of this system can be described by two state variables: the volume $V(t)$ of the water in the reservoir and $\Phi(t)$ the angular position of the see-saw, see fig. 1.2.1b. In the state plane the periodic action is represented by a closed curve.

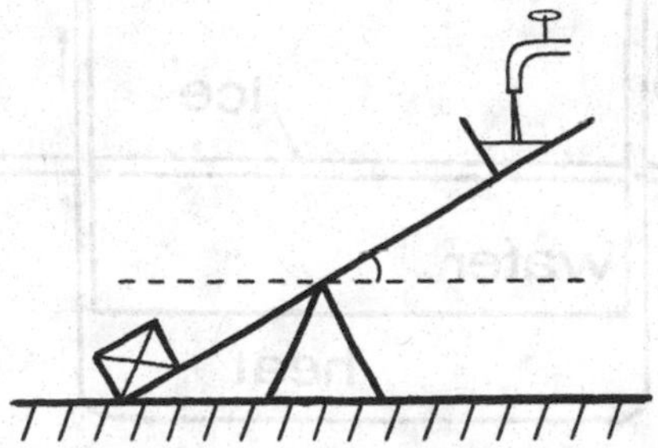

(a) construction of a see-saw with a water reservoir

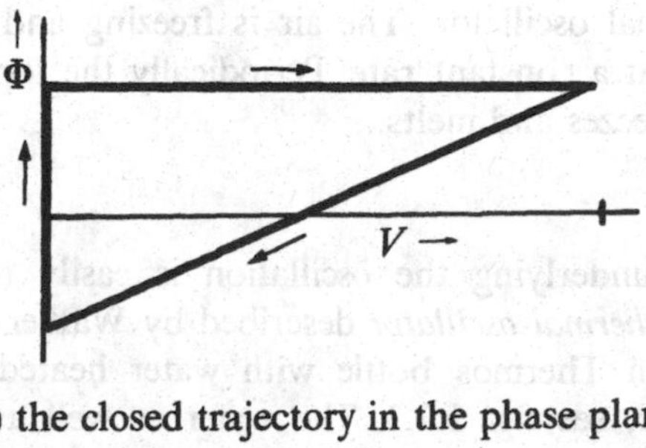

(b) the closed trajectory in the phase plane.

Fig.1.2.1 A mechanical relaxation oscillator: its state is determined by two variables the angular position Φ of the see-saw and the volume V of water in the reservoir.

The second mechanical device is a weight on a conveyor-belt, held in its position by a spring, see fig. 1.2.2. Assuming that there is some friction, we easily conclude that the position $x(t)$ changes in the same manner as the volume of water in the reservoir of the see-saw, see also Vatta (1979).

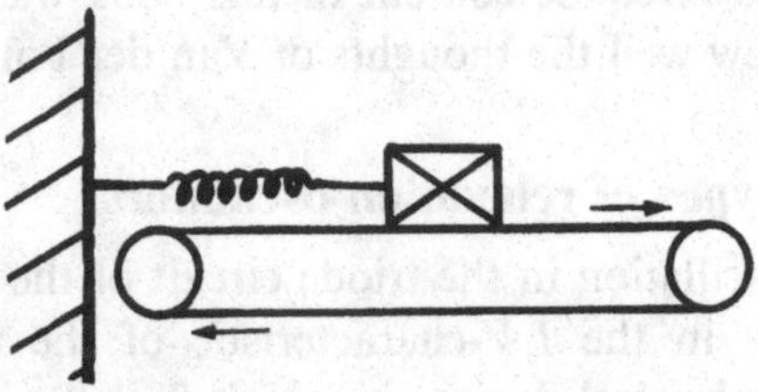

Fig.1.2.2 A mechanical relaxation oscillator: the conveyor-belt

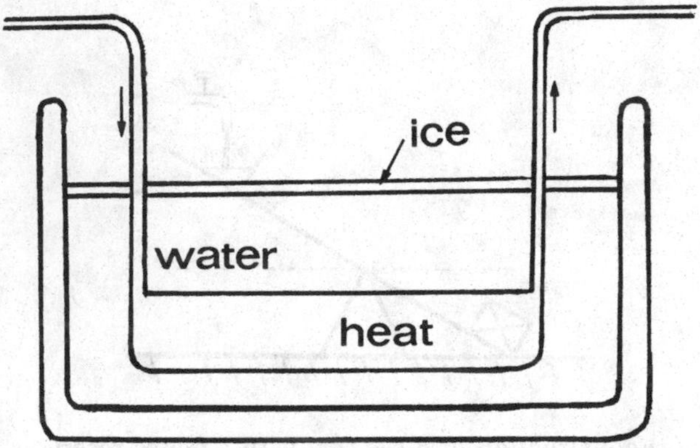

Fig.1.2.3 A thermal oscillator. The air is freezing and the water is heated at a constant rate. Periodically the top layer of the water freezes and melts.

The mechanism underlying the oscillation is easily recognized in other examples, such as a *thermal oscillator* described by Walker (1983). It works as follows. A wide open Thermos bottle with water heated by an element is placed in freezing air, see fig. 1.2.3. The water as well as the air above the water are stirred. Taking appropriate values for the temperature of the air and the heating capacity of the element, one observes the periodic freezing and melting of the top layer of the water. These oscillations are due to poor transfer of heat through ice.

1.3 Relaxation oscillations in physics and biology

Apart from the circuit of fig. 1.1.2 relaxation oscillations are present in many other types of electronic circuits, see Hayashi (1964) and Kaplan and Yafle (1980). New are circuits with superconducting elements, which may also give arise to relaxation oscillations. The SQUID (Superconducting Quantum Interference Device) is such an element. It consists of of a ring of superconducting metal interrupted by two *Josephson junctions,* being very thin insulating barriers, which allow a current to pass depending upon the magnetic field. The Josephson junction satisfies the equation

$$C\frac{dV}{dt} + \frac{V}{R} + I_c \sin\phi = I_e, \tag{1.3.1a}$$

where C is the junction capacitance, R the normal state resistance and ϕ and V, respectively, the difference between the phases of the quantum states and the potential difference over the junction. The maximum supercurrent through the junction is denoted by I_c and the external current by I_e. The Josephson relation

$$\frac{d\phi}{dt} = \frac{2e}{h}V \quad (h\,/\,2e \text{ is the flux quantum}) \tag{1.3.1b}$$

completes the system of differential equations that describes the dynamics of a Josephson junction. Eqs. (1.3.1) are equivalent to

$$\frac{d^2\phi}{dt^2} + \sigma\frac{d\phi}{dt} + \omega_0^2 \sin\phi = I\,, \tag{1.3.2}$$

where

$$\sigma = \frac{1}{RC}, \quad \omega_0^2 = \frac{2e}{h}\frac{I_c}{C}.$$

Schlup (1974) postulates that this equation needs to be modified and suggests the phenomenological equation

$$\beta\frac{d^2\phi}{dt^2} + (1+\gamma\cos\phi)\frac{d\phi}{dt} + \sin\phi = \alpha, \tag{1.3.3}$$

which exhibits relaxation oscillations in some domain of the parameter space ($|\gamma|>1,\ 0<\beta\ll1,\ 0<\alpha<1$), see Schlup (1979, 1981) and Sanders (1983).

In geophysics there exist two periodic phenemona which typically have the dynamics of a relaxation oscillation; geysers and earthquakes (see Ito *et al.,* 1980). In both cases an amount of energy is stored within a characteristic relaxation time. When the threshold is reached (boiling of water, maximum strain at a fold), this energy is released. Characteristic for these oscillations is the fluctuation of the period, which is one of the properties of a relaxation oscillation, see sections 1.1 and 2.6.2. The geyser, called the "Old faithful", in the Yellowstone Park is an exception in this respect. Its name reflects the punctuality of its bursts of steam every 65 minutes. Fluctations in the climate (ice ages) can also be modeled by a relaxation oscillator, see North (1985). Its dynamics is gouverned by the phase transition of water in ice and visa versa like in the thermal oscillator of the preceding section.

In chemistry reaction schemes may allow of periodic fluctuations in the concentration of some of the reactants. The Belousov-Zhabotinskii reaction is the best studied example, see Tyson (1976). It consists of 11 reactions, which are reduced to 5 basic steps involving 5 reactants, from which 2 are kept at a constant concentration. Then this reaction is described by three variables being the varying concentrations. Rate constants, fixed concentrations and stoichiometric coefficient act as parameters which can be manipulated. Thus, the reaction satisfies a system of differential equations of the form

$$\frac{dx_i}{dt} = f_i(x_1,x_2,x_3;p_1,\dots,p_4), \quad i = 1,2,3. \tag{1.3.4}$$

We will discuss this problem in section 2.4.2. Stochastic biochemical oscillations are modeled by Ebeling *et al.* (1985).

In biology oscillatory processes play a key role in the temporal organization of activity in an organism. At the cellular level there is synchronization of neural and cardiac oscillators by cyclic inputs, as well as mutual synchronization, see Holden (1976). At the same level the dynamics of the energy metabolism may exihibits relaxation oscillations. At a higher organization level organs possess a cyclic action structure: the regular beat of the heart, the respiration of the lungs and periodic contraction waves of the intestines are three examples of such time periodic behaviour. The organism as a whole exhibits cyclic activity known as the circadian rhythm: the rest-activity cycle of about 24hrs. Members of a species may go well-tuned to each other through a life cycle (aggregation of slime mold amoebae), see Segel (1980). Interactions between species (prey-predator relation) may give rise to oscillations in the densities of these species, see section 2.3.

Since biological systems may have a complex structure, that is not understood in detail, one usually works with a simple model for an oscillating system. In our investigations of chapter 3 we take the minimal representation of a dynamical system with a periodic solution: it is assumed to be described completely by 2 state variables, see also Bojadziev (1984). To remain as close as possible to the formalism of classical mechanics, Johannesma (1984) proposes to write such a system as

$$\frac{dp}{dt} = -\frac{\partial C}{\partial q} + \frac{\partial D}{\partial p}, \tag{1.3.5a}$$

$$\frac{dq}{dt} = \frac{\partial C}{\partial p} + \frac{\partial D}{\partial q}, \tag{1.3.5b}$$

where $C(p,q)$ and $D(p,q)$ are called the conservation and dissipation function. For $D \equiv 0$ Eq.(1.3.5) is a Hamiltonian system and $C(p,q)$ the conserved quantity. Biological systems are essentially dissipative ($D \not\equiv 0$). However, a Hamiltonian system may be a good first approximation, like the harmonic oscillator is for mechanical and electrical oscillations, e.g. in section 2.3 we deal with a model of population cycles being a Hamiltonian system. For $C \equiv 0$ Eq. (1.3.5) is a gradient system, which is used in catastrophe theory, see Thom (1971) and Zeeman (1977). Consequently, both lines of investigation of physical and biological systems can be picked up from the starting point (1.3.5). Another advantage of choosing this formalism will be clear when we consider coupled oscillators, see section 1.6. Any function $A(p,q)$ defined along a trajectory satisfies

$$\frac{dA}{dt} = \nabla A.\nabla D + [A,C], \tag{1.3.6}$$

where [,] denote the so-called Poisson brackets

$$[A,C] = \frac{\partial C}{\partial p}\frac{\partial A}{\partial q} - \frac{\partial C}{\partial q}\frac{\partial A}{\partial p}. \tag{1.3.7}$$

Choosing $A = C$ and $A = D$ we obtain the evolution equations for the conservation and dissipation function

$$\frac{dC}{dt} = \nabla C.\nabla D, \tag{1.3.8a}$$

$$\frac{dD}{dt} = \nabla D.\nabla D + [D,C]. \tag{1.3.8b}$$

It is noted that ΔD equals the divergence of the vector field of (1.3.5) and that the system is dissipative for $\Delta D < 0$.

1.4 Discontinuous approximations

The asymptotic behaviour of a singularly perturbed oscillator is easily understood from the heavily damped harmonic oscillator

$$\frac{d^2x}{d\tau^2} + \nu\frac{dx}{d\tau} + x = 0, \quad \nu >> 1 \tag{1.4.1}$$

or

$$\epsilon\frac{dx}{dt} = y - x, \tag{1.4.2a}$$

$$\frac{dy}{dt} = -x, \tag{1.4.2b}$$

with $\epsilon = 1/\nu^2 << 1$ and $t = \tau/\nu$.

Let the initial values be

$$x(0) = a, \quad y(0) = b \quad (a < b). \tag{1.4.3}$$

For $\epsilon = 0$ we obtain the *reduced system*

$$0 = y_0 - x_0, \tag{1.4.4a}$$

$$\frac{dy_0}{dt} = -x_0 \tag{1.4.4b}$$

or

$$x_0(t) = y_0(t) = C_0 e^{-t}. \tag{1.4.5}$$

Obviously, this approximate solution of (1.4.2) does not satisfy the initial values (1.4.3), while moreover the value of C_0 is still undetermined. Since initially the right-hand side of (1.4.2a) is positive, the derivative of x is of order $O(1/\epsilon)$ and so x rapidly increases until the line $y=x$ is reached. During this short period of time, y remains almost unchanged. Consequently, in the limit $\epsilon \to 0$ the solution jumps at $t=0$ from the initial point (a,b) to the value (b,b), see fig.1.4.1. Now we are in the position to determine C_0 in the approximate solution (1.4.5) valid for $t>0$:

$$C_0 = b. \tag{1.4.6}$$

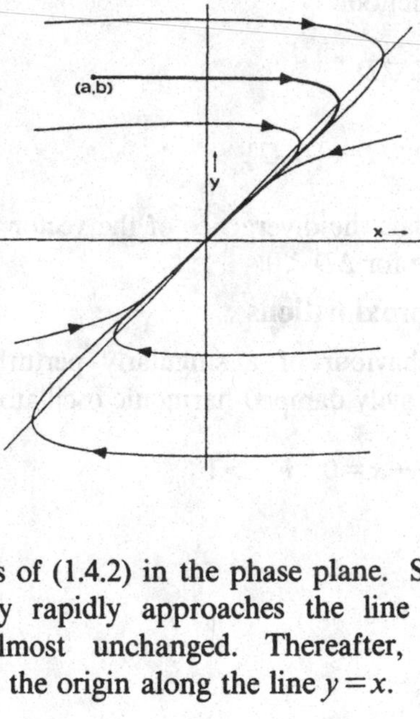

Fig.1.4.1 Trajectories of (1.4.2) in the phase plane. Starting in (a,b) a trajectory rapidly approaches the line $y=x$, while y remains almost unchanged. Thereafter, the trajectory approaches the origin along the line $y=x$.

In the examples of the preceding sections we observed that also for relaxation oscillations some of the variables may vary rapidly during a short time interval and that others fluctuate regularly. Let the vector function $x(t)$ represent the fast variables and $y(t)$ the slow ones. Then such a system can be described by the differential equations

$$\epsilon \frac{dx}{dt} = F(x,y), \tag{1.4.7a}$$

$$\frac{dy}{dt} = G(x,y) \tag{1.4.7b}$$

with $0<\epsilon<<1$. The discontinuous approximation satisfies

$$F(x,y) = 0. \tag{1.4.8}$$

The point, where the solution leaves this manifold $\mathcal{F}$ in state space, and the point, where it returns infinitely rapidly, depend upon F and G and in more exceptional cases also upon the initial values.

Clearly, Eq. (1.4.8) must be a nonlinear equation in x, because for y fixed it must have at least two roots: one corresponds with the leaving point (usually a double root) and the other with the return point. For y fixed the roots of Eq. (1.4.8) are the stationary points of Eq. (1.4.7a). When these are (un)stable the manifold $\mathcal{F}$ is called locally (un)stable.

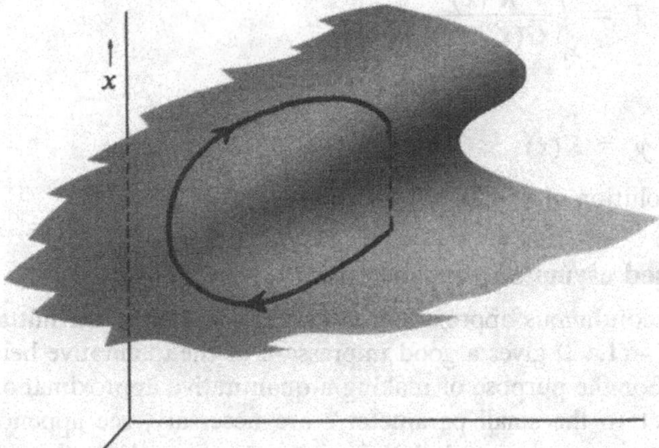

Fig.1.4.2 Relaxation oscillation in a system with one fast and two slow variables. For most of the time the trajectory of the periodic solution remains close to the manifold $\mathcal{F}$. At a point where $\partial F/\partial x=0$ the manifold $\mathcal{F}$ is left and the solution rapidly tends to a stable point of $\mathcal{F}$.

Let us consider an example with one "fast" variable and two "slow" variables. Expressing x as a function of y we obtain from (1.4.8)

$$x = H(y). \tag{1.4.9}$$

Since for y fixed Eq. (1.4.8) may be satisfied by different values of x, expression (1.4.9) is locally valid and one may have to switch from one branch to another. In fig. 1.4.2 we sketch a manifold $\mathfrak{F}$ that is stable except for the middle branch at the position where this manifold is folded. For any starting value away from $\mathfrak{F}$, the solution of (1.4.1) will jump instantaneously (in the limit for $\epsilon \to 0$) to one of the stable branches. The differential equation

$$\frac{dy}{dt} = G(H(y),y) \tag{1.4.10}$$

is assumed to have a periodic solution of period

$$T = \oint \frac{1}{G(H(y),y)} ds \tag{1.4.11}$$

The function $H(y)$ has different representations during the cycle, see fig. 1.4.2. At $y = y_0$ there is a jump from one branch to a next one. In order to compute the period it may be advantageous to integrate over x:

$$T = \int_{x_1}^{x_2} \frac{K'(x)}{G(x,K(x))} dx, \tag{1.4.12}$$

where

$$y = K(x) \tag{1.4.13}$$

is a local solution of (1.4.8).

1.5 Matched asymptotic expansions

The discontinuous approximation (1.4.3) - (1.4.6) of the initial value problem (1.4.2) - (1.4.3) gives a good impression of the qualitative behaviour of the solution. For the purpose of making a quantitative approximation, expansions with respect to the small parameter ϵ are necessary, see appendix B. In the previous section we noticed that initially the time derivative of x is of the order $O(1/\epsilon)$. Thus, x changes $O(1)$ in a time interval of order $O(\epsilon)$.

To analyse the change at that time scale we introduce a *stretching* transformation

$$\tau = t/\epsilon, \tag{1.5.1}$$

so that $\tau = O(1)$ for $t = O(\epsilon)$. The system (1.4.2) transforms into

$$\frac{dx}{d\tau} = y - x, \quad x(0) = a, \tag{1.5.2a}$$

$$\frac{dy}{d\tau} = -\epsilon x, \quad y(0) = b. \tag{1.5.2b}$$

In fluid mechanics such a stretching is applied in the variable normal to the boundary of an object in a flow. The stretched region is called the *boundary layer*.

It is assumed that the solution of (1.5.2) may be expanded in a power series with respect to ϵ:

$$x(\tau;\epsilon) = u_0(\tau) + \epsilon\, u_1(\tau) + \cdots, \tag{1.5.3a}$$

$$y(\tau;\epsilon) = v_0(\tau) + \epsilon\, v_1(\tau) + \cdots \tag{1.5.3b}$$

This is called the inner expansion (within the boundary layer). Substitution in (1.5.2) yields after equating the coefficients of equal powers in ϵ a recurrent system of equations for u_i and v_i:

$$\frac{du_0}{d\tau} = v_0 - u_0, \quad u_0(0) = a, \tag{1.5.4a}$$

$$\frac{dv_0}{d\tau} = 0, \quad v_0(0) = b, \tag{1.5.4b}$$

$$\frac{du_i}{d\tau} = v_i - u_i \quad ,u_i(0) = 0, \tag{1.5.5a}$$

$$\frac{dv_i}{d\tau} = -u_{i-1} \quad ,v_i(0) = 0 \tag{1.5.5b}$$

with $i = 1,2,\ldots$. The solutions of the first two equations read

$$u_0 = (a-b)e^{-\tau} + b, \quad v_0 = b,$$

$$u_1 = \int_0^\tau v_1(\sigma)e^{-(\tau-\sigma)}d\sigma, \tag{1.5.6a}$$

$$v_1 = (a-b)(e^{-\tau}-1) - b\tau. \tag{1.5.6b}$$

For τ tending to infinity t leaves the ϵ-neigbourhood of $t=0$ and the solution approaches the line $y=x$, where the regular approximation (1.4.5) holds. This asymptotic solution also improves if a power series expansion in ϵ is made: the so-called outer expansion,

$$x(t;\epsilon) = x_0(t) + \epsilon x_1(t) + \cdots, \tag{1.5.7}$$

$$y(t;\epsilon) = y_0(t) + \epsilon y_1(t) + \cdots,$$

where x_0 and y_0 satisfy (1.4.4) - (1.4.5). The recurrent system of equations for the other coefficients is as follows

$$y_i - x_i = x'_{i-1}(t), \tag{1.5.8a}$$

$$\frac{dy_i}{dt} = -x_i. \tag{1.5.8b}$$

For $i=1$ the solution is

$$x_1 = -y'_1(t), \quad y_1 = C_1 e^{-t} - C_0 t e^{-t}. \tag{1.5.9}$$

It is noted that both expansions (1.5.3) and (1.5.7) have a limited time interval over which they are valid. That is an interval where they converge asymptotically. Since u_1 and v_1 are of order $O(\tau)$ for τ large, the asymptotic expansion

(1.5.3) converges asymptotically for $t=\epsilon\tau=o(1)$. For the same reason (1.5.7) converges asymptotically for $\epsilon t=O(1)$.

We now come to the point of determining the constants C_i. This is done by matching (1.5.7) with (1.5.3). This can be carried out in different ways. Matching is possible because of the fact that the time intervals where both expansions converge asymptotically have overlap. Let us stretch the time variable in a way different from (1.5.1):

$$\sigma = t/\delta(\epsilon) = \tau\epsilon/\delta(\epsilon). \tag{1.5.10}$$

Then we may again consider the solution of (1.4.2) as a function of σ and the parameter ϵ. If we choose $\delta(\epsilon)$ such that for $\sigma=O(1)$ we are in the domain of overlap, expansions (1.5.3) and (1.5.7) should be identical and so C_0 and C_1 can be determined. Since τ must be large and t small, we have the condition

$$\delta(\epsilon) = o(1) \quad\text{and}\quad \epsilon/\delta(\epsilon) = o(1). \tag{1.5.11}$$

The computations can be slightly simplified by substituting $t=\tau\epsilon$ in (1.5.3) and requiring identical asymptotic behaviour of both expansions for τ large. Thus, Eq. (1.5.7) transforms into

$$x = C_0 e^{-\tau\epsilon} + \epsilon(C_1 + C_0 - C_0\tau\epsilon)e^{-\tau\epsilon} + \cdots, \tag{1.5.12}$$

$$y = C_0 e^{-\tau\epsilon} + \epsilon(C_1 - C_0\tau\epsilon)e^{-\tau\epsilon} + \cdots,$$

or

$$x = C_0 + \epsilon(-C_0\tau + C_0 + C_1) + O(\epsilon^2), \tag{1.5.13}$$

$$y = C_0 + \epsilon(-C_0\tau + C_1) + O(\epsilon^2).$$

On the other hand for τ large

$$u_0 \approx b, \quad u_1 \approx -b\tau + 2b - a, \tag{1.5.14}$$

$$v_0 \approx b, \quad v_1 \approx -b\tau + b - a,$$

so that identical expansions are found for

$$C_0 = b \quad\text{and}\quad C_1 = b - a. \tag{1.5.15}$$

The use of the asymptotic expansions as an approximation of the solution in case ϵ has a fixed small positive value brings about the problem of dividing the time interval in two separate subintervals: one near $t=0$ where (1.5.3) holds and the other away from from $t=0$ (but bounded independent of ϵ). This problem is overcome by the construction of a composite expansion, which is valid in any interval $[0,L]$ independent of ϵ. By adding the two expansions (1.5.3) and (1.5.7) and subtracting the matching terms (1.5.14) we obtain the so-called composite expansion, which is valid in both the time intervals,

$$x(t;\epsilon) = \{x_0(t) + u_0(t/\epsilon) - b\} +$$

$$+\epsilon\{x_1(t) + u_1(t/\epsilon) + b\tau - 2b + a\} + O(\epsilon^2)$$

$$y(t;\epsilon) = \{y_0(t)+v_0(t/\epsilon)-b\}+\epsilon\{y_1(t)+v_1(t/\epsilon)+b\tau-b+a\}+O(\epsilon^2)$$

or

$$x(t;\epsilon) = be^{-t}+(a-b)e^{-t/\epsilon}+O(\epsilon), \tag{1.5.16a}$$

$$y(t;\epsilon) = be^{-t}+O(\epsilon). \tag{1.5.16b}$$

This result is easily verified by expansion of the exact solution of (1.4.2) - (1.4.3).

For systems of type (1.4.7) similar locally valid asymptotic expansions can be constructed. It may occur that besides t the variables x and y have to be stretched in order to analyse the solution at a critical point in state space (a transition layer). In the next chapter the asymptotic analysis of relaxation oscillations of the type (1.4.7) forms the main topic. Special attention is given to the Van der Pol oscillator.

1.6 Forced oscillations

We already noted that the period of a relaxation oscillation is sensitive to external perturbations. Therefore, we will pay attention to stochasticly perturbed oscillations in section 2.6. However, fluctuation of the period may also be caused by an internal mechanism. In an oscillator described by three components, a strange attractor may be responsible for the irregularity of the oscillation. In section 2.6 we will deal with this phenomenon and construct a chaotic Van der Pol type relaxation oscillator, see also Gollub *et al.* (1980) and Herzel and Ebeling (1985).

Relaxation oscillations easily get entrained to periodic inputs. Van der Pol and Van der Mark (1927) investigated this phenomenon almost immediately after the discovery of the free relaxation oscillations, see section 1.1. They analyzed the frequency of the circuit of fig. 1.6.1a as a function of the capacitance C. In fig. 1.6.1b it is easily seen that the electrical system takes a period being a multiple of the forcing period. It is also observed that for certain parameter values two different subharmonics may coexist. Furthermore, there are regions where no subharmonic could be detected. From our asymptotic analysis of section 4.3.3 we conclude that this is a transient phenomenon: in such a region for some time, solutions stay, close to a chaotic solution before locking in. Chaotic solutions of the forced Van der Pol relaxation oscillator have been analyzed by Anand (1983), Levi (1981) and Grasman *et al.* (1984). The use of a telephone receiver, by which subharmonic solutions are detected from their tone, probably stimulated Van der Pol to study consonance of tones mathematically. In a paper on this subject, see Van der Pol (1946), he uses series of Farey fractions to identify musical scales. Recently, this line of investigation has been picked up by Yoshizawa *et al.* (1982) and Allen (1983).

Forced oscillations are of fundamental importance in electric and electronic devices, where signals need a perfect tuning. Again we refer to Hayashi (1964) and to applications of Josephson junctions, see Sanders (1983), Crutchfield *et*

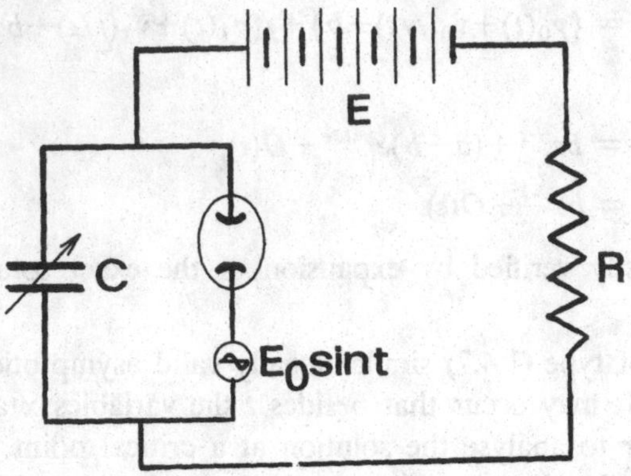

(a) the circuit

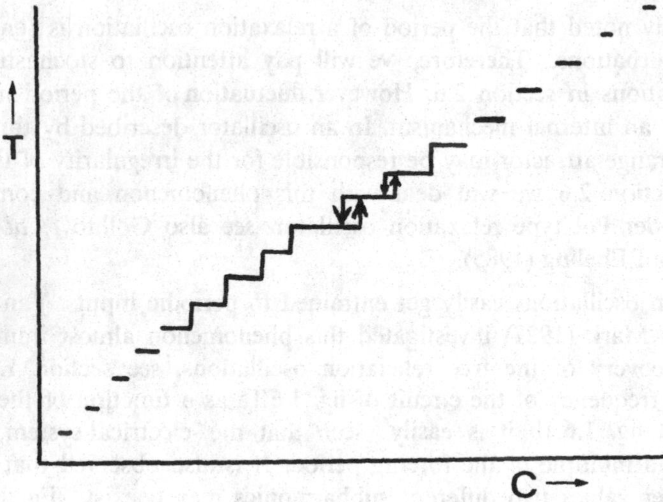

(b) existence of subharmonic solutions and transient phenomena for different values of *C*. The hysteresis phenomenon indicated by the arrows is due to the coexistence of different subharmonics.

Fig.1.6.1 Forced oscillation in an electronic circuit designed by Van der Pol and Van der Mark *al.* (1980) and Matisoo (1980).

In biology most organisms have some mechanism that synchronizes and adapts internal activities to cyclic changes outside, such as the circadian rhythm, which is regulated by the external light-dark cycle (Wever, 1979). The heart, driven by the sinus node, is one more example of a forced oscillation taking place in an organism, see Guevara *et al.* (1981) and Guevara and Glass (1982). Pacemakers are present in the process of aggregation of slime mold amoebae (Segel, 1980) and as a final example of external periodic forcing we mention the influence of the seasons upon population cycles, see Blom *et al.* (1981).

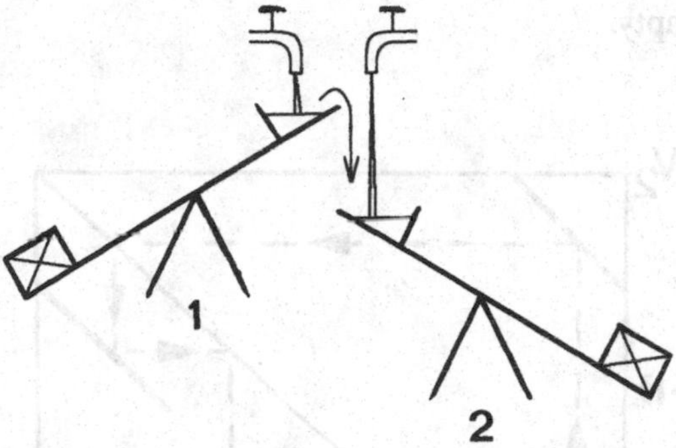

Fig.1.6.2 Two see-saws. When 1 turns over, it empties itself in the reservoir of 2. The sizes of the reservoirs W_i and periods T_i, needed to fill them, are such that $W_1 < W_2$ and $T_1 < T_2$.

The system of two see-saws of fig. 1.6.2 illustrates properties of a forced relaxation oscillator. The reservoir of 1 is smaller than that of 2. Oscillator 2 is said to be in its *refractory phase* if it is not completely filled when 1 turns over. If it is the case oscillator 2 is *triggered* by 1. The *phase* of each oscillator may be defined as the volume of water $V_i(t)$ in case at time $t=0$ the reservoirs are empty and the see-saws are put separate for $t \geqslant 0$. Let the free running periods be such that $T_1 < T_2$. In fig. 1.6.3 we give the phases in case oscillator 1 influences 2, as sketched in fig 1.6.2 It is assumed that at $t=0$ both reservoirs are empty. For general starting values the dynamics is analyzed from the functional relation between $V_2(T_1)$ and $V_2(0)$ given $V_1(0)=0$, see fig. 1.6.4. Let $V_2^{(k)} = V_2(kT_1)$, then we have the iteration map

$$V_2^{(k+1)} = PV_2^{(k)}$$

with P given in fig. 1.6.4.

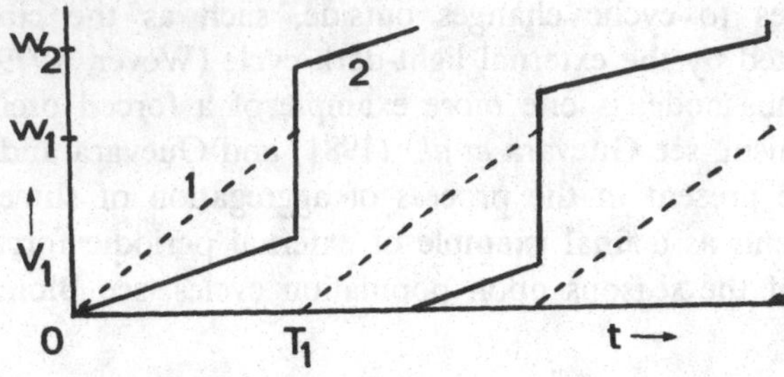

Fig.1.6.3 The volume in each reservoir as a function of time t when see-saw 1 empties itself in 2. Initially both reservoirs are empty.

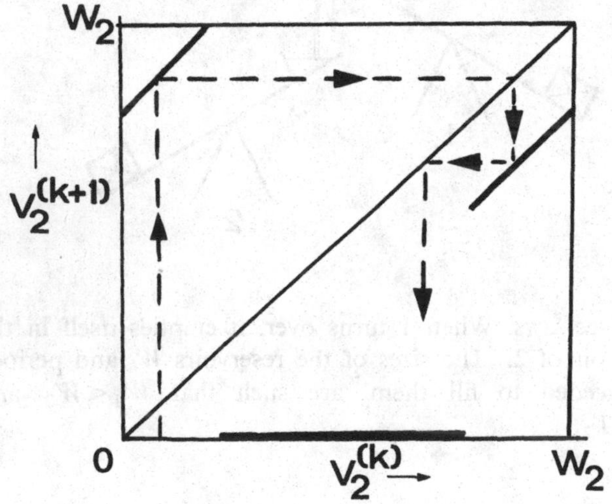

Fig.1.6.4 The iteration map P. Taking a starting value $V_2^{(0)}$ we obtain the iterates $V_2^{(k)}, k=1,2,...$ by following the dotted lines.

1.7 Mutual entrainment

The collective flashing of the southeast Asian fireflies is probably the most instructive example of the mechanism of mutual entrainment in biological systems, see Hanson (1978). Winfree (1967) developed a framework for describing the process of mutual entrainment in populations of coupled oscillators. Some elements of his model are used in our analysis of coupled relaxation oscillators, see chapter 3. Mutual entrainment of two oscillators with slightly different autonomous frequencies leads to a common rhythm with the inherently faster oscillators running ahead. For oscillators with widely different autonomous frequencies the spectrum of actual frequencies will exhibit peaks which are spaced in such a way that their positions have a ratio $n:m$ with n and m integers. This result may explain the presence of peaks in the spectrum of coupled neural oscillators. The occurrence of two entrained oscillations in an electronic circuit is analyzed by Gollub *et al.* (1978) for the circuit given in Fig. 1.7.1a. The two tunnel diodes have characteristics as sketched in Fig 1.7.1b. For this circuit the voltage and current satisfy the system of differential equations:

$$C_1 dV_1/dt = I_1 - F(V_1), \tag{1.7.1a}$$

$$L_1 dI_1/dt = E - V_1 - R(I_1 + I_2), \tag{1.7.1b}$$

$$C_2 dV_2/dt = I_2 - F(V_2), \tag{1.7.1c}$$

$$L_2 dI_2/dt = E - V_2 - R(I_1 + I_2). \tag{1.7.1d}$$

For R, C_1 and C_2 of small order of magnitude this system is of the type we will study with asymptotic methods.

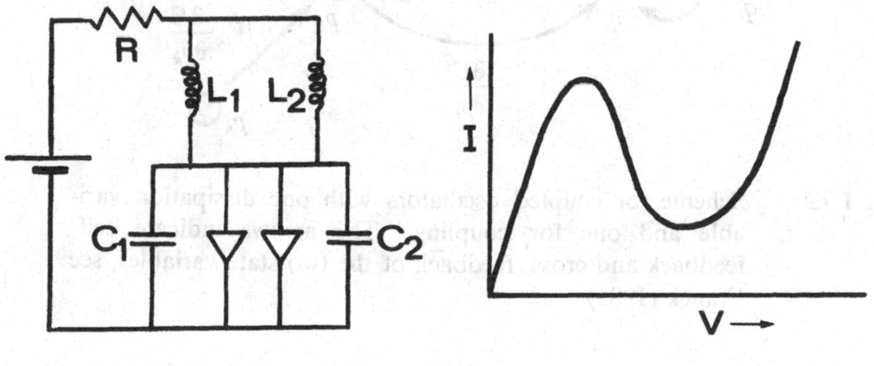

(a) the circuit (b) characteristic of the tunnel diode

Fig.1.7.1 Oscillations in an electronic network with two tunnel diodes.

A second phenomenon due to mutual entrainment is phase-wave propagation. In systems of spatially distributed oscillators with coupling to the nearest neighbour, entrained solutions occur in which the phases of the oscillators depend upon time and position of the oscillator in such a way that a travelling wave is formed. The contraction waves of the intestines is an example of 1-dimensional wave propagation. Fibrillation of the heart muscle can be seen as an irregular wave pattern in 2 or 3- dimensions.

Assuming that there is a diffusion type of coupling between the oscillators, we obtain, in the limit of the spatial discretization of oscillators tending to zero, a reaction-diffusion system. Especially for chemical systems, but also for spatially distributed prey-predator systems, the existence of phase waves has been analyzed from these nonlinear diffusion equations, see section 3.4.6. The reader is also referred to Fife (1977), Chow and Tam (1976), Bonilla and Liñán (1984) and Kuramoto (1984).

Returning to the C,D-system (1.3.5), we make the following assumption for a system of coupled oscillators: one component of the oscillator regulates the interaction with the environment (dissipation), while the other one forms the link with the other oscillators. This decomposition is sketched in fig. 1.7.2.

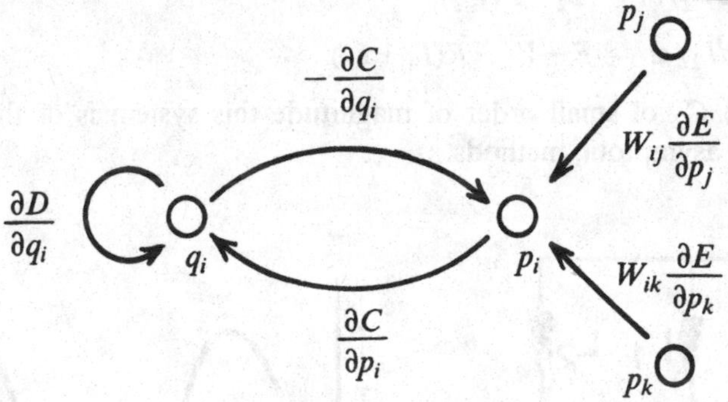

Fig.1.7.2 Scheme for coupled oscillators with one dissipation variable and one for coupling. The arrows indicate self-feedback and cross feedback of the two state variables, see Franck (1985).

The internal structure remains unchanged and is given by the conservation function C for the system of coupled oscillators:

$$C = \sum_i C_i(p_i,q_i), \quad D = \sum_i D_i(q_i). \tag{1.7.2}$$

Moreover, the interaction between the oscillators is assumed to consist of two elements: the message function

$$E(p,q) = \sum_i E_i(p_i,q_i)$$

and the transfer function $W_{ij}(p_i,p_j)$. Consequently, a system of coupled oscillators is described by

$$\frac{dp_i}{dt} = -\frac{\partial C}{\partial q_i} + \sum_j W_{ij}\frac{\partial E}{\partial p_j}, \tag{1.7.3a}$$

$$\frac{dq_i}{dt} = \frac{\partial C}{\partial p_i} + \frac{\partial D}{\partial q_i}. \tag{1.7.3b}$$

The evolution equations for conservation and dissipation are

$$\frac{dC}{dt} = \nabla_q C.\nabla_q D + \nabla_p C.W\nabla_p E,$$

$$\frac{dD}{dt} = [C,D] + \nabla_q D \nabla_q D.$$

Example. For biological systems diffusion of a reaction component forms a mechanism of interaction between oscillators. Thus, in the model system (1.7.3) the message function consists of the components p_i:

$$E = \sum_i p_i,$$

while transfer by diffusion takes the form

$$W_{ij} = a_{ij}(p_j - p_i),$$

where a_{ij} depends upon the spatial structure of the system of coupled oscillators, e.g. $a_{ij} = 0$ except for neighbouring oscillators, where it has a fixed positive value.

Exercises

1.1. Let the tunnel diode of the circuit given below have an I,V-characteristic satisfying

$$I = F(V - V_0).$$

Derive the second order differential equation that gouverns this system.

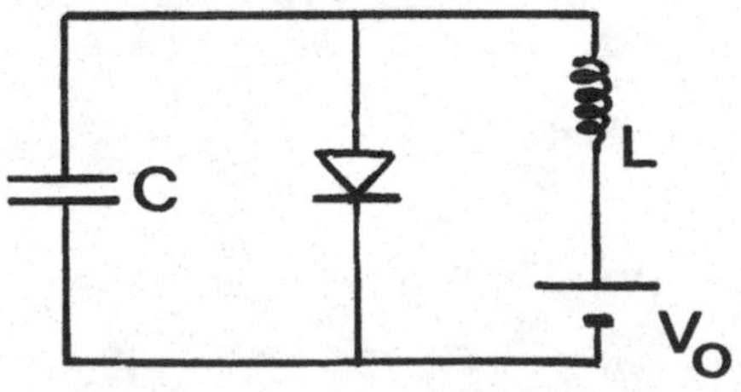

1.2 Compute in two ways the period of the discontinuous approximation of the periodic solution of the singularly perturbed system

$$\epsilon\frac{dx}{dt} = y - F(x),$$

$$\frac{dy}{dt} = -x,$$

with

$$F(x) = \begin{cases} x-2 & \text{for } x>1, \\ -x & \text{for } |x|<1, \\ x+2 & \text{for } x<-1. \end{cases}$$

1.3 Derive the differential equation for the circuit of fig. 1.5.1a.
1.4 Give the conservation and dissipation function for the harmonic oscillator in the representation (1.3.5). Give these functions also for the system of exercise 1.2.
1.5 Derive the evolution equation for the message function $E(p,q)$ given by (1.6.2).
1.6 The Van der Pol equation (1.1.3) is for $0<\nu<<1$ approximated by

$$x = A \cos t.$$

compute A from the requirement that the conservation function is periodic (the right-hand side of (1.3.8a) has to vanish after averaging over one period).
1.7 Investigate the following system of two coupled relaxation oscillators.

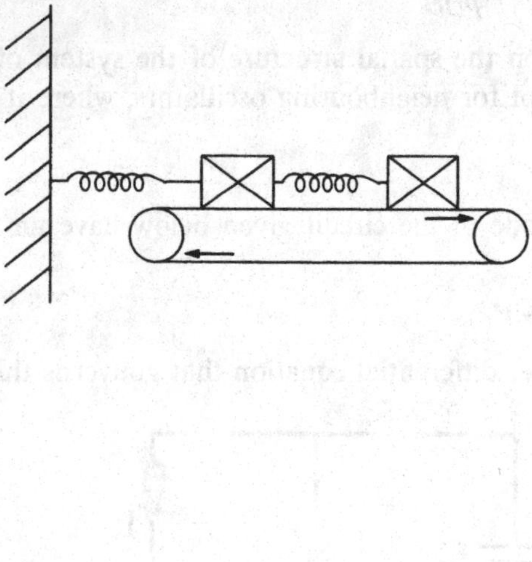

2. FREE OSCILLATION

The definition of a relaxation oscillation is presented. A review of the proofs of existence of periodic solutions of singularly perturbed systems is given. One such a method, based upon the extension theorem, is worked out. The different asymptotic solutions of the Van der Pol oscillator are given in detail. A similar asymptotic analysis of the Volterra-Lotka equations is made. We deal with Van der Pol oscillators with a stochastic and a constant forcing term and finally construct a chaotic relaxation oscillator.

2.1 AUTONOMOUS RELAXATION OSCILLATION: DEFINITION AND EXISTENCE

In this section we will formulate a definition of relaxation oscillations. In the introduction a notion of relaxation oscillations evolved, which is more or less a generalization of the Van der Pol oscillator. The trajectory of this oscillator follows in state space, for most of the time, a stable manifold. At a fold the trajectory leaves the manifold and rapidly returns to it at a point of a different branch. The dynamics of such systems are also studied in relation with catastrophe theory, see Zeeman (1977). When the trajectory closes, the system describes a relaxation oscillation. The mathematical analysis of this generalized Van der Pol oscillator is found in the book by Mishchenko and Rosov (1980), which brings together results from Mishchenko (1961), Pontryagin (1961) and Mishchenko and Pontryagin (1960). An asymptotic approximation theorem from Tikhonov (1948) forms the starting point of their investigations, see section 2.1.4. This approach is also followed by Nipp (1983).

In section 2.1.1 we give a phenomenological definition of a relaxation oscillation. Two elements constitute this definition: the dependence of the solution upon a parameter of the system, and, secondly, the rapid change in some of the state variables during a short time interval of the cycle. By this definition nonlinear oscillations of a type different from the generalized Van der Pol oscillator are also characterized as relaxation oscillations. As an example we mention periodic solutions of the Volterra-Lotka equations in a certain range of parameter values, see section 2.3. In this way the definition differs slightly from the generally accepted notion of relaxation oscillations, see Cronin (1977). At the other hand we feel that we stay remarkably close to Van der Pol's idea of relaxation oscillations as we discussed in section 1.1.

For the study of oscillations in a two dimensional state space the Poincaré-Bendixson theory can be used, see section 2.1.2.

In section 2.1.3 we take a different position and use a technique based upon the extension theorem. The idea behind it was formulated heuristicly by Kaplun (1957) for boundary layer problems in fluid mechanics. Eckhaus (1979) gave this method its mathematical foundation in the extension theorem. In Eckhaus (1983) a global description of its application to relaxation oscillations is given. In section 2.1.3 this is worked out in more detail.

A third approach of analyzing mathematically relaxation oscillations has its roots in nonstandard analysis, see Callot *et al.* (1978). Diener (1980), Bélair (1983) and Bélair and Holmes (1984). It is felt that strong parallels exist between this approach and the previous one. An important point is that these two methods are not restricted to generalized Van der Pol type relaxation oscillators.

2.1.1. A mathematical characterization of relaxation oscillations

The dynamics of the triode circuit of section 1.1 can be described by two state variables, x and $dx/d\tau$. We rewrite Eq.(1.1.3) as a system of two coupled first order differential equations in these state variables, let $p = dx/d\tau$, then

$$\frac{dx}{d\tau} = p, \tag{2.1.1a}$$

$$\frac{dp}{d\tau} = -\nu(x^2-1)p - x. \tag{2.1.1b}$$

In the Lienard representation one defines

$$y = \frac{dx}{d\tau} + \nu(\frac{1}{3}x^3 - x),$$

so that the system takes the form

$$\frac{dx}{d\tau} = y - \nu(\frac{1}{3}x^3 - x), \tag{2.1.2a}$$

$$\frac{dy}{d\tau} = -x. \tag{2.1.2b}$$

Substitution of $x = -dy/d\tau$ in Eq. (2.1.2a) yields, after scaling te y variable, the so-called Rayleigh equation

$$\frac{d^2y}{d\tau^2} - \nu\{\frac{dy}{d\tau} - (\frac{dy}{d\tau})^3\} + y = 0,$$

which has been analyzed by Lord Rayleigh (1883) for ν small.

In general, a physical system can be described by n state variables satisfying a set of n coupled first order differential equations. When moreover at a time, say $\tau = 0$, the values of the n state variables are known, then the values at all times can be computed from the differential equations with initial values. We write such a system as

$$\frac{dx_i}{d\tau} = f(x_1,..,x_n;\nu), \quad x_i(0) = x_{i0}, \quad i = 1,...,n. \tag{2.1.3}$$

It is supposed that the parameter ν of this system represents the time constant related to the period in case the system exhibits a relaxation oscillation. Since these oscillations are present for $\nu >> 1$, it is more convenient to change the time scale and the relaxation parameter:

$$\tau = t\nu, \quad \epsilon = 1/\nu. \tag{2.1.4}$$

We now write (2.1.3) as

$$p_i(\epsilon)\frac{dx_i}{dt} = h_i(x_1,...,x_n;\epsilon), \quad i = 1,...,n, \tag{2.1.5}$$

where p_i and h_i are continuous functions. Moreover, it is assumed that p_i and h_i have bounded limit values as $\epsilon \to 0$ and that for $i(j)$, $j = 1,...,k$, $k \geq 1$, $p_{i(j)}$

vanishes as ϵ tends to zero. In state space a periodic solution is represented by a closed curve L_ϵ.

Definition 2.1.1 A curve M in state space is called a strict transversal, if it is transversal to all L_ϵ, $0<\epsilon\leqslant\epsilon_0$ (ϵ_0 fixed), and, if moreover, an arbitrarily small positive number δ independent of ϵ exists, such that the acute angle θ_ϵ between M and L_ϵ satisfies $|\theta_\epsilon|\geqslant\delta>0$.

Definition 2.1.2 A periodic solution of (2.1.5) is called a **relaxation oscillation** if its period $T(\epsilon)$ has a bounded nonzero limit value as $\epsilon\to 0$ and, if moreover, two strict transversals M_f and M_s exist, such that

$$\lim_{\epsilon\to 0}\frac{p_i(\epsilon)}{h_i(x_\epsilon;\epsilon)} = 0 \text{ for some } i \text{ with } x_\epsilon = M_f\cap L_\epsilon \tag{2.1.6a}$$

and

$$\lim_{\epsilon\to 0}\frac{h_i(x_\epsilon;\epsilon)}{p_i(\epsilon)} \text{ is bounded for all } i \text{ with } x_\epsilon = M_s\cap L_\epsilon. \tag{2.1.6b}$$

It is remarked that only nonlinear systems of the type (2.1.5) may exhibit a relaxation oscillation. Furthermore, this definition does not answer the question of stability of the oscillation. At this point our definition is not in agreement with the existing idea that relaxation oscillations are self-sustained. The essential property of a relaxation oscillation is its fast change during a short time interval of the cycle.

Lienard (1928) studied systems of the type

$$\epsilon\frac{dx}{dt} = y - F(x), \tag{2.1.7a}$$

$$\frac{dy}{dt} = -x \tag{2.1.7b}$$

with $F(x)$ such that the system has a relaxation oscillation. Clearly, the Van der Pol equation with $F(x)=1/3x^3-x$ belongs to this class. In the Lienard plane the trajectories satisfy

$$\{y-F(x)\}dy/dx=-\epsilon x. \tag{2.1.8}$$

For x bounded independenly of ϵ, we have approximately

$$\{y-F(x)\}\ dy/dx=0. \tag{2.1.9}$$

Thus for $\epsilon\to 0$ the trajectories either tend to a line y = constant or to the curve $y=F(x)$, see fig.2.1.1. According to (2.1.7b) the period satisfies

$$T=\oint-\frac{dy}{x}. \tag{2.1.10}$$

The contribution over the line segement AB is

$$\int_{x_A}^{x_B} -\frac{dy}{x} = \int_{x_A}^{x_B} -\frac{dy}{dx}\frac{dx}{x} = \int_{x_A}^{x_B} \frac{\epsilon dx}{y-F(x)} = O(\epsilon). \qquad (2.1.11)$$

This is the short time interval where x changes rapidly. For the segment BC we obtain

$$\int_{y_B}^{y_C} \frac{dy}{x} = \int_{y_B}^{y_C} \frac{dF(x)}{x} = \int_{x_B}^{x_C} \frac{F'(x)}{x} dx. \qquad (2.1.12)$$

This yields together with the segment DA the first order approximation to the period. For the Van der Pol equation we find $T(0)=3-2\ln 2$. More detailed computations have been given by Haag (1943, 1944).

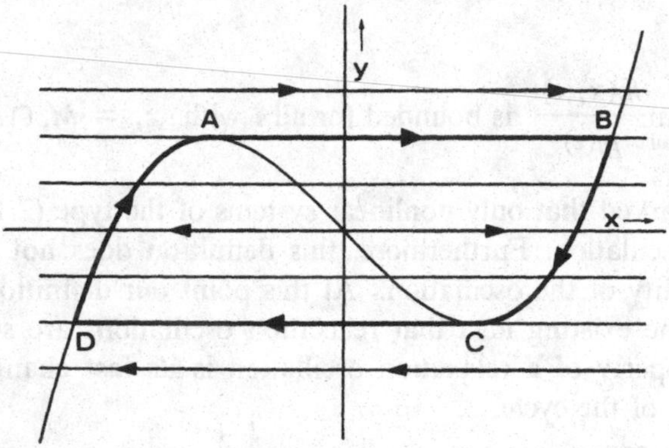

Fig.2.1.1 Trajectories of the system (2.1.7) in the limit case $\epsilon \to 0$. For any starting value, the origin excluded, the solution approaches the closed trajectory ABCD.

2.1.2 Application of the Poincaré-Bendixson theorem

There is a topological method to prove the existence of periodic solutions of two dimensional nonlinear systems of the type

$$\frac{dx_1}{dt} = f_1(x_1, x_2), \quad \frac{dx_2}{dt} = f_2(x_1, x_2). \qquad (2.1.13)$$

We consider trajectories which for $t \geq 0$ remain in a ring shaped domain Ω of the phase plane. This domain is bounded by two closed curves Γ_1 and Γ_2, see fig.2.1.2. The Poincaré-Bendixson theory deals with the set of limit points of such a trajectory.

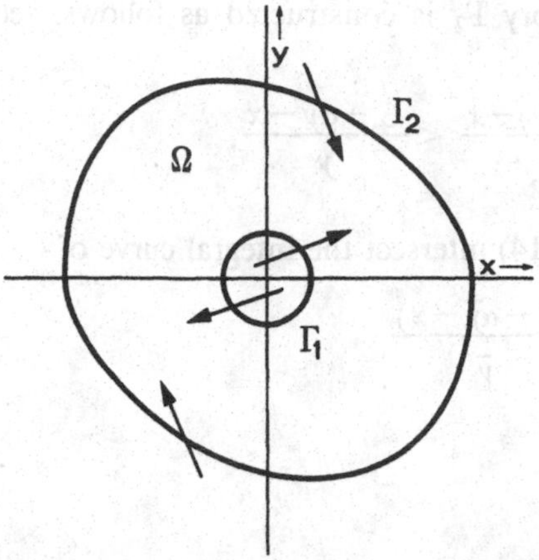

Fig.2.1.2 The domain Ω in the phase plane. All trajectories that intersect $\partial\Omega = \Gamma_1 \cup \Gamma_2$ are entering Ω for increasing t. The domain Ω does not contain a stationary point of (2.1.13).

Theorem 2.1.1 If a trajectory of (2.1.13) remains in Ω for $t \geqslant 0$ and Ω does not contain an equilibrium point of (2.1.13), then the trajectory or its set of limit-points forms a limit cycle.

Stoker (1950) uses this theorem for proving the existence of periodic solutions of systems of the type

$$\frac{dx}{dt} = y, \quad \frac{dy}{dt} = F(y) - x \tag{2.1.14}$$

satisfying the following conditions

$$F(y) = G(y) - \alpha y, \tag{2.1.15}$$

with

$$G(y) = -G(-y), \quad |G(y)| < C, \quad G'(y) > 0, \quad G'(0) > \alpha, \tag{2.1.16}$$

where α and C are positive constants. The linearized equation takes the form

$$\frac{dx_0}{dt} = y_0, \quad \frac{dy_0}{dt} = \beta y_0 - x_0, \quad \beta = F'(0) > 0. \tag{2.1.17}$$

Clearly, the linearized system is unstable. From Lyapunov's stability theory it follows that $(x,y) = (0,0)$ is also an unstable equilibrium of the nonlinear system (2.1.14). Consequently, a circle exists, with center in the origin, such that all trajectories of (2.1.14) intersecting this circle are directed outwards. A second closed trajectory Γ_2 is constructed as follows, see fig.2.1.2. Since for $y>0$

$$\frac{dy}{dx}=\frac{F(y)-x}{y}<\frac{C-\alpha y-x}{y}, \tag{2.1.18}$$

all trajectories of (2.1.14) intersect the integral curve of

$$\frac{d\bar{y}}{dx}=\frac{(C-\alpha\bar{y}-x)}{\bar{y}} \tag{2.1.19}$$

and have

$$\frac{dy}{dx}<\frac{d\bar{y}}{dx}$$

in the point of intersection. Consequently, for increasing t all trajectories are directed inwards at Γ_2. Let us check this for $\alpha<2$ and introduce $q^2=1-1/4\alpha^2$. Integration of (2.1.19) yields

$$\bar{y}(t)=-Ae^{-1/2\alpha t}\sin qt, \tag{2.1.20a}$$

$$x(t)=C+Ae^{-1/2t}(\frac{1}{2}\sin qt+q\cos qt), \tag{2.1.20b}$$

where t is a parameter and A an integration constant with a positive value. Moreover, the integral curve is chosen such that $\bar{y}(0)=0$. For $t=-\pi/q$ the function $\bar{y}(t)$ vanishes again. Now we choose A such that

$$x(-\pi/q)=-x(0),$$

so

$$A=\frac{2C}{q}\{\exp(\frac{\alpha\pi}{2q})-1\}^{-1}. \tag{2.1.21}$$

The curve Γ_2 is completed by taking $(x,\bar{y})=(-x,-\bar{y})$ for $\bar{y}<0$. Since all trajectories enter Ω and Ω does not contain an equilibrium point, the system has a periodic solution with a closed trajectory contained in Ω.

In a more refined appoach, the domain Ω has the shape of a narrow ring enclosing the limit cycle, so that the proof of existence is also constructive in the sense that an approximation of the periodic solution is found. This method has been used for relaxation oscillations by Levinson and Smith (1942), Flanders and Stoker (1946), LaSalle (1949) and Ponzo and Wax (1965 abc).

We follow LaSalle, who takes the system (2.1.7), for some class of functions $F(x)$. As it is our goal to convey the idea of his method, we confine ourselves to the more restricted class of odd, continuously differentiable functions $F(x)$ satisfying

$$F'(x)<0 \quad \text{for } |x|<c, \tag{2.1.22a}$$
$$F'(x)>0 \quad \text{for } |x|>c. \tag{2.1.22b}$$

The limit equation (2.1.9) suggests that the limit cycle will be close to the curve $y=F(x)$ for $c \leqslant |x| \leqslant b$ and the lines $y=F(\pm c)$ connecting the two stable parts of the curve, see fig.2.1.3. For increasing t the limit cycle is taken clockwise. LaSalle constructs two closed curves Γ_1 and Γ_2 forming the inner and outer boundary of the domain Ω. The curve Γ_1 consists of the segments:
A_1B_1 satisfying

$$y = F(b_1) + \epsilon\sqrt{b_1^2 - x^2}$$

with $A_1=(0,F(-c))$ and $B_1=(b_1,F(b_1))$ (the constant b_1 is chosen such that the tangent of the line A_1B_1 equals $-\epsilon$),
B_1C_1 satisfying $y=F(x)$ with $C_1=(c,F(c))$,
C_1D_1 satisfying $y=F(c)$ with $D_1=(0,F(c))$,
D_1E_1, E_1F_1 and F_1A_1 are obtained from reflection of the above three segments with respect to the origin.

The curve Γ_2 consists of the segments:
A_2B_2 satisfying $y=F(b_2)$ with $A_2=(0,F(b_2))$ and $B_2=(b_2,F(b_2))$ (the constant b_2 is such that the tangent of A_1B_2 equals ϵ),
B_2C_2 being $y=F(x)-\epsilon\sqrt{b_2^2-x^2}$ with $C_2=(c,F(c)-\epsilon\sqrt{b_2^2-c^2})$,
C_2D_2 satisfying $y=F(c)-\epsilon\sqrt{b_2^2-x^2}$ with $D_2=(0,F(c)-\epsilon b_2)$,
D_2E_2, E_2F_2 and F_2A_2 are the reflections of these segments with respect to the origin. Notice that $F(b_2)=-F(c)+\epsilon b_2$.

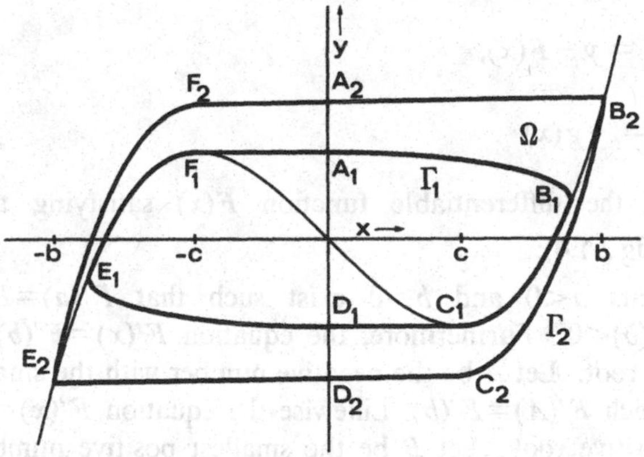

Fig.2.1.3 The construction of LaSalle for the domain Ω

It is easily verified that the domain Ω, enclosed by Γ_1 and Γ_2, does not contain an equilibrium point of the system. To verify that all trajectories enter Ω, we define for a segment $y=g(x)$ the normal vector $n=(g'(x),-1)$. For displacement along a trajectory we introduce the vector

$$v=(\frac{dx}{dt},\frac{dy}{dt})=(\frac{y-F(x)}{\epsilon},-x). \tag{2.1.23}$$

From the sign of the inner product (n,v) and the direction of the normal vector we conclude that indeed all trajectories, intersecting Γ_1 and Γ_2, enter the domain Ω. As an example we carry out the computations for the segment B_2C_2 of the curve Γ_2. This segment satisfies

$$g(x)=F(x)-\epsilon\sqrt{b_2^2-x^2}, \tag{2.1.24}$$

so that

$$(n,v)=g'(x)\{g(x)-F(x)\}+\epsilon^2 x \tag{2.1.25}$$

or

$$(n,v)=\epsilon\sqrt{b_2^2-x^2}\,\{-f(x)+\frac{x\epsilon}{\sqrt{b_2^2-x^2}}\}+\epsilon^2 x. \tag{2.1.26}$$

The vector n points outwards. Hence, v directs to the inside of Ω, as (n,v) is negative for ϵ sufficiently small. Consequently, a limit cycle L exists and Γ_1,Γ_2 tend to L as $\epsilon\to 0$.

Ponzo and Wax (1965 abc) analyse the system

$$\epsilon\frac{dx}{dt}=y-F(x), \tag{2.1.27a}$$

$$\frac{dy}{dt}=-g(x) \tag{2.1.27b}$$

with $g(x)$ and the differentiable function $F(x)$ satisfying the following conditions (see fig.2.1.4):

(a) Two constants $a<0$ and $b>0$ exist such that $F'(a)=F'(b)=0$ and $F'(a)>0, F'(b)<0$. Furthermore, the equation $F'(x)=F'(b)$ has at least one negative root. Let A be the negative number with the smallest absolute value for which $F'(A)=F'(b)$. Likewise the equation $F'(x)=F'(a)$ has at least one positive root. Let B be the smallest positive number for which $F(B)=F(a)$.

(b) An $\epsilon>0$ exists such that for $A-\epsilon\leqslant x\leqslant B+\epsilon$ the functions $g(x)$ and $F''(x)$ are both Lipschitz continuous and $xg(x)>0$.

(c) For the interval $a\leqslant x\leqslant b$ the functions $g'(x)$ and $F'''(x)$ are Lipschitz continuous.

(d) Two positive contstants K_1 and K_2 exist such that

$$F'(x)>K_1(a-x) \text{ for } A-\epsilon\leqslant x\leqslant a$$

and

$$F'(x) > K_2(x-b) \text{ for } b < x \leq B+\epsilon.$$

(e) Positive constants E_1, E_2, E_3 and E_4 can be found such that

$$E_2(a-x)^2 \leq F(a)-F(x) \leq E_1(a-x)^2, \tag{2.1.28}$$

$$E_4(x-b)^2 \leq F(x)-F(b) \leq E_3(b-x)^2 \tag{2.1.29}$$

in the interval $a \leq x \leq b$.

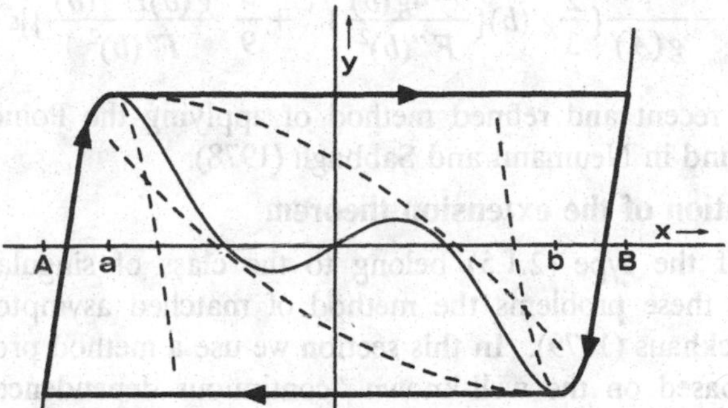

Fig.2.1.4 The null curve $y=F(x)$ in the method of Ponzo and Wax

Under the above conditions, the system (2.1.27) has a stable periodic solution, which is approximated by Ponzo and Wax (1965a) through the construction of a ring shaped domain Ω with width $O(\epsilon^{2/3})$ satisfying the conditions for the Poincaré-Bendixson theorem. In a next study the limit cycle L whithin Ω is approximated asymptotically and the following formula's for the positive and negative amplitude of x and the period are found, see Ponzo and Wax (1965b).

$$Ampl^+ = B + \frac{\alpha}{F'(B)}\left[\frac{-2g(a)^2}{F''(a)}\right]^{\frac{1}{3}}\epsilon^{\frac{2}{3}} + \tag{2.1.30}$$

$$+\frac{1}{2F'(B)}\left[\frac{2g(B)}{F'(B)}+\frac{2}{3}g'(a)\left\{\frac{-4g(a)}{F''(a)^2}\right\}^{\frac{1}{3}}+\right.$$

$$\left.+\frac{4}{9}\frac{g(a)F'''(a)}{F''(a)^2}\right]\epsilon\ln\epsilon+O(\epsilon),$$

where $-\alpha$ is the first zero of the Airy function, $\alpha = 2.33811$. The negative amplitude is found from replacing B by A and a by b. The time needed to go from $x = Ampl^-$ to $x = Ampl^+$ is

$$T = -\int_A^a \frac{F'(x)}{g(x)} dx + \alpha[\{\frac{2}{g(a)F''(a)}\}^{\frac{1}{3}} + \tag{2.1.31}$$

$$+ \frac{1}{g(A)}\{\frac{-2g(b)^2}{F''(b)}\}^{\frac{1}{3}}]\epsilon^{\frac{2}{3}} +$$

$$-\frac{1}{2}[\frac{2}{F'(B)} - \frac{2}{F'(A)} - \frac{2}{3}\frac{g'(a)}{g(a)F''(a)} + \frac{4}{9}\frac{F'''(a)}{F''(a)^2} +$$

$$-\frac{1}{g(A)}\{\frac{2}{3}g'(b)\{\frac{-4g(b)}{F''(b)^2}\}^{\frac{1}{3}} + \frac{4}{9}\frac{g(b)F'''(b)}{F''(b)^2}\}]\epsilon \ln \epsilon + O(\epsilon).$$

The most recent and refined method of applying the Poincaré-Bendixson theorem is found in Neumann and Sabbagh (1978).

2.1.3 Application of the extension theorem

Systems of the type (2.1.5) belong to the class of singular perturbation problems. To these problems the method of matched asymptotic expansions applies, see Eckhaus (1979). In this section we use a method proposed by Eckhaus (1982) based on the well-known "continuous dependence on the data" theorem and the "extension" theorem, which was first employed heuristicly by Kaplun (1954). This theorem appears to be of fundamental importance for proving the validity of matching procedures in singular perturbation theory.

Let the vector function $x(t,\epsilon)$ satisfy

$$\frac{dx}{dt} = f(x,t,\epsilon), \quad x(0,\epsilon) = x_0, \tag{2.1.32}$$

where f is continuous and uniformly bounded for $0 < \epsilon \leqslant \epsilon_0$ in a open bounded domain D of $\mathbb{R}^n$. Moreover, f is supposed to be Lipschitz-continuous with respect to x. Let us consider two such initial value problems

$$\frac{dx^{(i)}}{dt} = f_i(x^{(i)},t,\epsilon), \tag{2.1.33}$$

$$x^{(i)}(0,\epsilon) = x_0^{(i)}, \quad i = 1,2 \tag{2.1.34}$$

with $x_0^{(i)} \in D_0$, where $D_0 \subset D$ is a compact subset such that the distance between ∂D and ∂D_0 is bounded away from zero with a lower bound independent of ϵ. Furthermore, it is supposed that

$$|x_0^{(1)} - x_0^{(2)}| \leqslant \delta_0(\epsilon), \quad |f_1 - f_2| \leqslant \delta_f(\epsilon) \tag{2.1.35}$$

for all $x \in \overline{D}$ and $0 \leqslant t \leqslant K$ with $\delta_0(\epsilon)$ and $\delta_f(\epsilon)$ of order $o(1)$. The following theorem can be proved, see Eckhaus (1979).

Theorem 2.1.2 Let (2.1.33)-(2.1.34) with $i = 2$ have a solution which for $0 \leqslant t \leqslant L < K$ remains in D_0. If (2.1.35) is satisfied, then (2.1.33)-(2.1.34) with $i = 1$ has a solution for $0 \leqslant t \leqslant L$ with

$$|x^{(1)}(t,\epsilon) - x^{(2)}(t,\epsilon)| = O(\delta_0(\epsilon)) + O(\delta_f(\epsilon)). \tag{2.1.36}$$

Next we state a version of the extension theorem, which applies in particular to relaxation oscillations, see Eckhaus (1979) for the theorem in its general form.

Theorem 2.1.3 Let for $0 \leqslant t \leqslant L$ with L arbitrary

$$|x^{(1)}(t,\epsilon) - x^{(2)}(t,\epsilon)| = O(\delta), \quad \delta = o(1) \tag{2.1.37}$$

Then order functions $\delta_e(\epsilon)$ and $\tilde{\delta}(\epsilon)$ exist such that

$$|x^{(1)}(t,\epsilon) - x^{(2)}(t,\epsilon)| = O(\tilde{\delta}) \tag{2.1.38}$$

for $0 \leqslant t \leqslant \delta_e^{-1}(\epsilon)$ with $\tilde{\delta}(\epsilon)$ and $\delta_e(\epsilon)$ of order $o(1)$. If (2.1.37) holds for $0 < \eta \leqslant t \leqslant L$ with η arbitrary small, then order functions $\delta_e(\epsilon)$ and $\tilde{\delta}(\epsilon)$ can be found such that (2.1.38) holds for $0 < \delta_e(\epsilon) \leqslant t \leqslant L$.

The idea behind the extension method is to construct a locally valid approximation. By the extension theorem its validity is extended forward in the time domain. In the new domain an approximation is made and its domain of validity is again extended forward. One proceeds in this way until, inspite of repeated extensions, a time domain cannot be penetrated. Then an approximation is constructed in this domain and a backward extension is made. Since in the domain of overlap both approximations are valid, integration constants can be determined from matching techniques.

Example We will apply this rigorous method of matched local solutions to the scalar function $x(t,\epsilon)$ satisfying

$$\epsilon\frac{dx}{dt} = x - \tfrac{1}{3}x^3 - t, \quad x(0) = x_0 > \sqrt{3}. \tag{2.1.39}$$

This model problem, proposed by Nipp (1980), already possesses the characteristics of a relaxation oscillation.

The initial boundary layer. There is a boundary layer at $t = 0$, where x approaches the value $\sqrt{3}$. Since this layer is of order $O(\epsilon)$, we introduce the small time scale

$$\tau = t/\epsilon, \tag{2.1.40}$$

so that (2.1.39) transforms into

$$\frac{dx}{d\tau} = x - \tfrac{1}{3}x^3 - \tau\epsilon. \tag{2.1.41}$$

Using theorem 2.1.2 we obtain

$$|x(\tau)-\hat{x}(\tau)| = O(\epsilon) \quad \text{for } 0 \leq \tau \leq L \tag{2.1.42}$$

with

$$\frac{d\hat{x}}{d\tau} = \hat{x}-\frac{1}{3}\hat{x}^3, \quad \hat{x}(0) = x_0. \tag{2.1.43}$$

From theorem 2.1.3 it follows that

$$|x(\tau)-\hat{x}(\tau)| = o(1) \quad \text{for } 0 \leq \tau \leq \tau_1(\epsilon) \tag{2.1.44}$$

with $\tau_1(\epsilon) = L/\delta(\epsilon)$, where $\delta(\epsilon) = o(1)$. Since $\hat{x}$ tends to $\sqrt{3}$ as $\tau \to \infty$,

$$|x(\tau_1(\epsilon))-\sqrt{3}| = \sigma\epsilon = o(1). \tag{2.1.45}$$

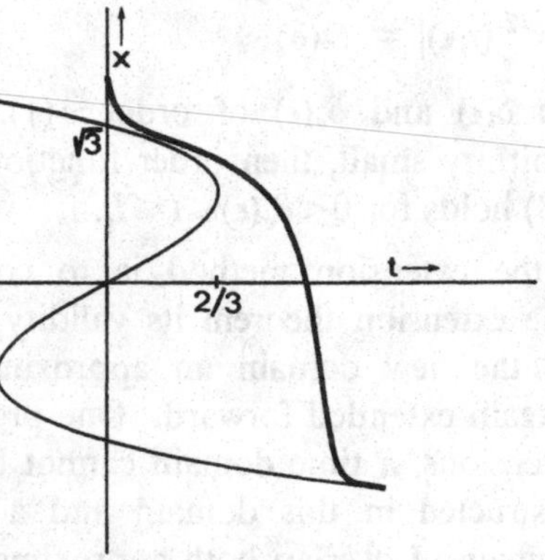

Fig.2.1.5 Asymptotic solution of the initial value problem (2.1.45)

The regular approximation. In fig.2.1.5 it is seen, that the solution approaches a stable regular solution $\tilde{x}(t;\epsilon)$ satisfying

$$\tilde{x}(t;\epsilon) = \sum_{k=0}^{\infty} \tilde{x}_k(t)\epsilon^k \tag{2.1.46}$$

with

$$\tilde{x}_0-\frac{1}{3}\tilde{x}_0^3-t = 0, \quad (1-\tilde{x}_0^2)\tilde{x}_1 = \frac{d\tilde{x}_0}{dt}, \ldots \tag{2.1.47}$$

The local behaviour of the solution near this curve is analyzed as follows. We substitute in (2.1.39)

$$x = \tilde{x}_0(t)+\epsilon\tilde{x}_1(t)+\sigma(\epsilon)\Phi, \tag{2.1.48}$$

$$t = \tau_1(\epsilon)+\tau\epsilon, \tag{2.1.49}$$

where $\sigma(\epsilon)$ is given by (2.1.45). This yields

$$\frac{d\Phi}{d\tau} = (1-\tilde{x}_{01}^2-\sigma\tilde{x}_{01}\Phi-\frac{1}{3}\sigma^2\Phi^2)\Phi +$$

$$+\frac{\epsilon^2}{\sigma}(-\frac{d\tilde{x}_1}{dt}+\tilde{x}_0\tilde{x}_1^2-\frac{1}{3}\epsilon\tilde{x}_1^3), \tag{2.1.50}$$

$$\Phi(0) = 1-\frac{\epsilon}{4\sigma} \tag{2.1.51}$$

with $\tilde{x}_{01} = \tilde{x}_0+\epsilon\tilde{x}_1$. The following estimate holds for $\tau\epsilon = o(1)$:

$$|\Phi(\tau)|\leqslant e^{-\tau}+O(\frac{\epsilon^2}{\sigma}\tau). \tag{2.1.52}$$

Thus, for $\tau = \tau_2(\epsilon) = O(\ln\epsilon^{-1})$ the solution is in an ϵ-neighbourhood of $\tilde{x}(t;\epsilon)$.

It is expected that it is approximated by $\tilde{x}(t;\epsilon)$ for

$$t\geqslant t_2 = \epsilon\tau_2(\epsilon) \tag{2.1.53}$$

until the asymptotic series (2.1.46) breaks down. From (2.1.47) it is seen that this is the case for x approaching the value 1. More precisely, the term

$$\epsilon\tilde{x}_1 = \epsilon t/(1-\tilde{x}_0^2)^2 \tag{2.1.54}$$

and all the higher order ones will be of order $O(\epsilon^{1/3})$ as $\tilde{x}$ enters an $\epsilon^{1/3}$-neighbourhood of the value 1.

The validity of the regular approximation can be proved by the method of Tikhonov, see Nipp (1983). One may use as well the approximation theorem 2.1.2 and the extension theorem 2.1.3. Then the following argument holds.

The stay near the regular solution. Suppose at some time $t_k\in(0,2/3)$ we have

$$x(t_k) = \tilde{x}(t_k)+O(\sigma_k), \quad \sigma_k = o(1). \tag{2.1.55}$$

Then by substituting

$$x = \tilde{x}+\sigma_k\Phi, \quad t = t_k+\epsilon\tau \tag{2.1.65}$$

in (2.1.39) and letting $\epsilon\to 0$, we find the approximating function $\hat{\Phi}$ satisfying

$$\frac{d\hat{\Phi}}{d\tau} = (1-\tilde{x}_0^2)\hat{\Phi}, \quad \hat{\Phi}(0) = 1. \tag{2.1.53}$$

Applying the extension theorem we conclude that at time $t_{k+1}>t_k$ an order function $\sigma_{k+1}(\epsilon)$ exists such that

$$x(t_{k+1}) = \tilde{x}(t_{k+1})+\sigma_{k+1}, \quad \sigma_{k+1} = o(1), \tag{2.1.58}$$

$$t_{k+1} = t_k+O(\epsilon/\delta), \quad \delta = o(1). \tag{2.1.59}$$

Starting at $t = t_2$ given by (2.1.57) we may continue until for some k

$$\tilde{x}_0^2(t_k)-1 = O(\delta_k), \quad \delta_k = o(1) \tag{2.1.60}$$

with

$$t_k = \tilde{x}_0-\frac{1}{3}\tilde{x}_0^3. \tag{2.1.61}$$

It is easily verified that indeed we arrive at such a point after $O(1/\epsilon)$ steps.

Then for an order function $\delta_k(\epsilon)$ satisfying

$$\epsilon/\delta_k^3 = o(1), \quad \sigma_k = o(\delta_k) \tag{2.1.62}$$

we introduce the transformations

$$t = t_k+\tau\epsilon/\delta_k, \quad x = \tilde{x}_0(t)+\delta_k\Phi. \tag{2.1.63}$$

Substitution in (2.1.39) yields

$$\frac{d\Phi}{d\tau} = -\Phi-\tilde{x}_0\Phi^2-\frac{1}{3}\Phi^3\delta_k-\tau\frac{\epsilon}{\delta_k^3}, \tag{2.1.64}$$

$$\Phi(0) = \sigma_k/\delta_k, \tag{2.1.65}$$

which is approximated by

$$\frac{d\hat{\Phi}}{d\tau} = -\hat{\Phi}-\hat{\Phi}^2, \quad \hat{\Phi}(0) = \sigma_k/\delta_k. \tag{2.1.66}$$

Consequently, a point $t_{k+1}>t_k$ exists with

$$x(t_{k+1}) = \tilde{x}_0(t_{k+1})+\sigma_{k+1}, \tag{2.1.67}$$

$$\tilde{x}_0^2(t_{k+1})-1 = O(\delta_{k+1}), \quad \delta_{k+1}/\delta_k = o(1) \tag{2.1.68}$$

Since

$$\sigma_{k+1} = o(\delta_k e^{-\delta_k/\delta_{k+1}}) \tag{2.1.69}$$

and

$$\delta_k/\delta_{k+1}e^{-\delta_k/\delta_{k+1}} = o(1), \tag{2.1.70}$$

we find that $\sigma_{k+1} = o(\delta_{k+1})$ and so we may proceed with the next step. From (2.1.64) it is seen that we are allowed to continue this procedure for the time that

$$\epsilon/\delta_k^3 = o(1).$$

Suppose an order function $\tilde{\delta}$ with $\epsilon/\tilde{\delta}^3 = o(1)$ exists such that $\delta_k \to \tilde{\delta}$ as $k \to \infty$. Then a local solution at $\tilde{\delta}$ of the type (2.1.63) is extended backward.

Next we carry out this procedure for the significant limit case $\tilde{\delta} = O(\epsilon^{\frac{1}{3}})$.

The transition layer. We introduce local variables v and τ satisfying

$$x = 1+v(\tau)\epsilon^{\frac{1}{3}}, \tag{2.1.71a}$$

$$t = \frac{2}{3}+\tau\epsilon^{\frac{2}{3}}. \tag{2.1.71b}$$

This transforms (2.1.39) into

$$\frac{dv}{d\tau} = -v^2-\frac{1}{3}v^3\epsilon^{-\frac{1}{3}}-\tau. \tag{2.1.72}$$

Each element of the set of solutions is approximated asymptotically by an element of the set of solutions of

$$\frac{d\hat{v}}{d\tau} = -\hat{v}^2-\tau. \tag{2.1.73}$$

Using the transformations $\hat{v} = z'(\tau)/z(\tau)$ we find for the set of solutions of (2.1.73):

$$\hat{v}(\tau) = \frac{Ai'(-\tau)+CBi'(-\tau)}{Ai(-\tau)+CBi(-\tau)}, \tag{2.1.74}$$

where $Ai(\cdot)$ and $Bi(\cdot)$ denote the Airy functions. By the extension theorem an order function $\delta = o(1)$ exists such that an element of $\{v(\tau)\}$ is approximated by an element of $\{\hat{v}(\tau)\}$ over a time interval $(-1/\delta(\epsilon),S)$ for some S. For $\tau\to-\infty$ there is overlap with the solutions of (2.1.63)-(2.1.65), so we require

$$\tilde{x}_o(t_k) = 1+\epsilon^{\frac{1}{3}}\hat{v}((t_k-\frac{2}{3})\epsilon^{-\frac{2}{3}})+o(\delta_k) \tag{2.1.75}$$

or

$$\epsilon^{\frac{1}{3}}\hat{v}((t_k-\frac{2}{3})\epsilon^{-\frac{2}{3}}) = \sqrt{\frac{2}{3}-t_k}+o(\sqrt{\frac{2}{3}-t_k}). \tag{2.1.76}$$

Only the element of $\{\hat{v}(\tau)\}$ with $C = 0$ satisfies this matching condition. Thus, upto now we approximated the solution for the time interval

$$t\in[0,2/3-(\alpha-\eta)\epsilon^{2/3}],$$

where $-\alpha$ is the first zero of the Airy function $Ai(\cdot)$ and η an arbitrarily small positive number. Applying the extension theorem once more, we obtain an approximation that holds for $t\in[0,t_0]$,

$$t_0 = \frac{2}{3}-\{\alpha-\frac{1}{\delta(\epsilon)}\}\epsilon^{\frac{2}{3}} \tag{2.1.77}$$

with $\delta = o(1)$ and

$$x(t_0)\approx 1-\epsilon^{\frac{1}{3}}/\delta(\epsilon). \tag{2.1.78}$$

Crossing the unstable region. In order to leave an $o(1)$-neighbourhood of $x = 1$ we continue as follows. Let

$$(x,t) = (x(t_k),t_k) \text{ with } t_k = \frac{2}{3}+O(\epsilon^{\frac{2}{3}})$$

and

$$x(t_k) = 1-\sigma_k(\epsilon), \quad \epsilon^{\frac{1}{3}}/\sigma_k = o(1). \tag{2.1.79}$$

Then using the transformation

$$t = t_k+\tau\frac{\epsilon}{\sigma_k}, \quad x = 1+\sigma_k\Phi, \tag{2.1.80}$$

we find

$$\frac{d\Phi}{d\tau} = -\Phi^2-\frac{1}{3}\sigma_k\Phi^3+-\tau\frac{\epsilon}{\sigma_k^3}+O(\frac{\epsilon^{2/3}}{\sigma_k^2}), \tag{2.1.81a}$$

$$\Phi(0) = -1. \tag{2.1.81b}$$

The approximation reads

$$\hat{\Phi} = 1/(\tau-1).$$

By the extension theorem the approximation holds for $0 \leqslant \tau \leqslant 1-\delta_{k+1}$. At $t = t_{k+1}$ we have

$$x(t_{k+1}) = 1+\sigma_{k+1}\Phi, \quad \sigma_k/\sigma_{k+1} = o(1), \tag{2.1.82a}$$

$$t_{k+1} = t_k+\epsilon(1-\delta_{k+1})/\sigma_k = \frac{2}{3}+O(\epsilon^{\frac{2}{3}}). \tag{2.1.82b}$$

Starting with $k = 0$ and $x(t_0)$ given by (2.1.77)-(2.1.78), we proceed until $\sigma_k = O(1)$, then a different approximation $\hat{\Phi}$ holds: the situation is identical to the initial boundary layer at $t = 0$, see (2.1.40)-(2.1.45). It can be shown that indeed only a finite number of steps are needed to have $\sigma_k = O(1)$. From the argument, we gave before for the sequence (2.1.63), it is concluded that an infinite set exists with $\sigma_k \to O(1)$ as $k \to \infty$. Extending backward the local approximation for the limit case, we will have overlap with a solution (2.1.80) for some k. Finally, we arrive in $x = -2$ at the second branch of the regular solution (2.1.46), see fig.2.1.5.

The generalized Van der Pol oscillator We continue from section 1.4 and consider the singularly perturbed system of the type

$$\epsilon\frac{dx_i}{dt} = F_i(x,y;\epsilon), \quad i = 1,...,m, \tag{2.1.83a}$$

$$\frac{dy_j}{dt} = G_j(x,y;\epsilon), \quad j = 1,...,n, \tag{2.1.83b}$$

where ϵ is a small positive parameter. For the points (x,y) at the manifold

$$\mathcal{F} = \{(x,y)|F(x,y) = 0\}$$

it is assumed that a subset $\mathbb{S}$ exists for which the matrix

$$A = \left[\frac{\partial F_i}{\partial x_j}\right] \qquad (2.1.84)$$

has eigenvalues with negative real parts: the stable manifold. At the boundary of $\mathbb{S}$ only one eigenvalue is assumed to have a real part behaving as a simple zero, then $\det(A) = 0$. This problem has been analyzed by Mishchenko and Pontryagin, see section 2.1.4. A proof based upon the extension theorem is along the same lines as for the example given above.

The initial boundary layer. For a starting value $(x_0, y_0) \notin \mathfrak{F}$ an initial boundary layer arises. The asymptotic approximation satisfies

$$\frac{d\tilde{x}}{d\tau} = F(\tilde{x}, \tilde{y}; 0), \quad \tilde{x}(0) = x_0, \qquad (2.1.85a)$$

$$\frac{d\tilde{y}}{d\tau} = 0, \tilde{y}(0) = y_0. \qquad (2.1.85b)$$

For $\tau \to \infty$ the solution approaches the manifold $\mathfrak{F}$ at a point of $\mathbb{S}$.

The regular expansion. Near $\mathbb{S}$ the solution is expanded as

$$x = \sum_{k=0}^{\infty} \tilde{x}_k(t)\epsilon^k, \quad y = \sum_{k=0}^{\infty} \tilde{y}_k(t)\epsilon^k. \qquad (2.1.86)$$

The coefficients $\tilde{x}_k$, and $\tilde{y}_k$ satisfy a recurrent system of equations that is found by substitution of (2.1.86) in (2.1.83).

Leaving the stable manifold. When we arrive at a leaving point p at the boundary of $\mathbb{S}$, the real part of one of the eigenvalues of A vanishes. Let the eigenvectors at this point be ξ_i, $i = 1, \ldots, m$ with ξ_{i_1} being the vector that corresponds with the critical eigenvalue. Moreover, let one variable $y_{j_1}(t)$ be monotonic near p. Then we may proceed as in the example with x replaced by ξ_{i_1} and t by y_{j_1}, see (2.1.71)-(2.1.82). The other variables remain constant in the local approximation in a similar way as y in (2.1.85). Once the solution has left a $o(1)$-neighbourhood of the point p, we handle as in the case of the initial boundary layer.

Periodicity. Let us assume that after some time a trajectory γ arrives in a $o(1)$-neighbourhood of a point of $\mathbb{S}$ (away from the boundary), where it has been before. Then we take a $(m+n-1)$-dimensional transversal intersection Σ and consider the Poincaré mapping

$$P: \Sigma \to \Sigma. \qquad (2.1.87)$$

When the flow is contracting within the manifold $\mathfrak{F}$ near the trajectory γ, then Brouwer's fixed point theorem applies: a compact set within Σ containing $\gamma \cap \Sigma$ is mapped into itself and so a fixed point corresponding with a periodic solution exists. The asymptotic approximation of the period is seen as a part of the formal computations and follows from the matched local solutions.

As we will learn in the next section, the system (2.1.83) may also have a solution of a chaotic type. The extension method also applies to solutions of that form.

Other types of relaxation oscillators. Our definition of a relaxation oscillation, see section 2.1.1, comprises other types oscillations than just the generalized Van der Pol oscillator.

We mention oscillations of a conservative system within some parameter range, see section 2.3, where we deal with the Volterra-Lotka equations. The validity of the asymptotic approximation of a trajectory can be proved with the extension theorem. Since such an approximation of the trajectory returns in a $o(1)$-neighbourhood of a point in the phase plane, it indeed yields an approximation of the closed orbit through this point.

For canard-type of oscillations, see section 2.5.2, the validity of the asymptotic approximation of trajectories can also be proved. Since the system stays in a plane, one may use the Poincaré-Bendixson theorem to prove the existence of a periodic solution.

2.1.4 Application of Tikhonov's theorem

In this section we give an overview of the results by Mishchenko (1961), Pontryagin (1961) and Mishchenko and Pontryagin (1960) for generalized Van der Pol type oscillators. For the complete review of their method of analysis we refer to the book of Mishchenko and Rosov (1980). In our description we modify the method at two points. First, our construction of the mappings π_i^ϵ of theorem 2.1.6 is more straightforward. Secondly we study the possibility of chaotic solutions.

Generalized Van der Pol type oscillators are gouverned by systems of differential equations of the type (2.1.83):

$$\epsilon \frac{dx}{dt} = F(x,y), \tag{2.1.88a}$$

$$\frac{dy}{dt} = G(x,y), \tag{2.1.88b}$$

where ϵ is a small positive parameter and x and y denote vector functions with m and n components. The functions F_i and G_j are assumed to be sufficiently many times differentiable. We also consider the corresponding reduced system

$$F(x,y) = 0, \tag{2.1.89a}$$

$$\frac{dy}{dt} = G(x,y). \tag{2.1.89b}$$

In the $(m+n)$-dimensional phase space Eq.(2.1.89a) represents the n-dimensional manifold $\mathfrak{F}$ see fig. 2.1.6. In points of $\mathfrak{F}$, where $\det(A) \neq 0$ with A

given by (2.1.84), we may solve (2.1.89a) with respect to x:

$$x = \chi(y). \tag{2.1.90}$$

Substitution in (2.1.89b) yields

$$\frac{dy}{dt} = G(\chi(y),y). \tag{2.1.91}$$

Thus, a solution of (2.1.89) takes the form

$$x = \chi(\psi(t)), \quad y = \psi(t). \tag{2.1.92}$$

This solution at $\mathcal{F}$ approximates a trajectory of the full system (2.1.88) at intervals of slow action. In the fast action away from $\mathcal{F}$, y is taken constant and x satisfies

$$\epsilon\frac{dx}{dt} = F(x,y). \tag{2.1.93}$$

The transition from slow action to fast action is at a *leaving point* of $\mathcal{F}$, where $\det(A) = 0$. When the solution of (2.1.93) approaches a stable equilibrium the trajectory returns in a slow action.

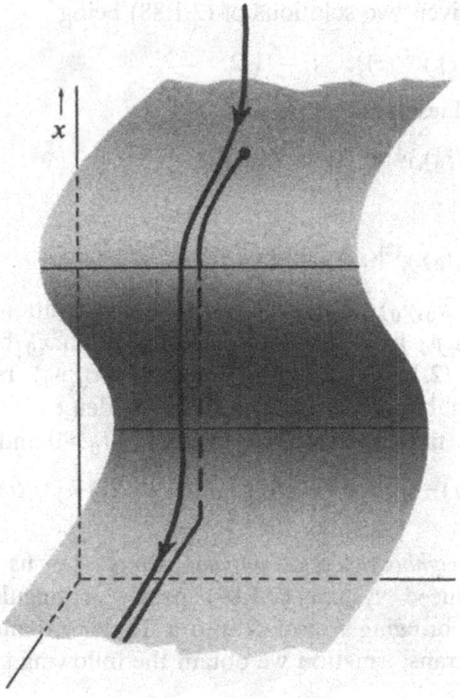

Fig.2.1.6 The manifold $\mathcal{F}$ with the stable and unstable branches separated by the sets of points, where $\det(A)=0$.

Behaviour outside a neighbourhood of singular points. In phase space $\mathbb{R}^{m+n}$ away from singular points situated at $\mathcal{F}$, we analyse the trajectories of (2.1.88) by the following two approximation theorems.

Theorem 2.1.4 Let $\{x_0(t), y_0(t)\}$ be a solution of the reduced system (2.1.89), which for $t \in [t_0, t_1]$ is contained in $\mathcal{F}$ where the eigenvalues of A have negative real parts bounded away from zero. Then any solution of (2.1.88) with initial values

$$|x(t_0;\epsilon) - x_0(t_0)| = O(\epsilon), \quad |y(t_0;\epsilon) - y_0(t_0)| = O(\epsilon) \tag{2.1.94}$$

satisfies

$$x(t;\epsilon) = x_0(t) + O(\epsilon), \quad y(t;\epsilon) = y_0(t) + O(\epsilon) \tag{2.1.95}$$

for $t \in [t_0, t_1]$.

This result was established by Tikhonov (1952). For a refinement of this theorem extending the time domain in case of relaxation oscillators we refer to Nipp (1983). A generalization of the theorem is proved by Siška and Dvořak (1984). In a second theorem it is described how two solutions of (2.1.88) relate if one of them starts in $\mathcal{F}$.

Theorem 2.1.5 Given two solutions of (2.1.88) being

$$\{x^{(i)}(t), y^{(i)}(t)\}, \quad i = 1,2.$$

Let the initial values satisfy

$$\{x^{(1)}(t_0), y^{(1)}(t_0)\} = (x_0, y_0) \tag{2.1.96a}$$

and

$$\{x^{(2)}(t_0), y^{(2)}(t_0)\} = \{\overline{x}_0, y_0\} \tag{2.1.96b}$$

with $(x_0, y_0) \in \mathcal{F}$, $(\overline{x}_0, y_0) \notin \mathcal{F}$ and $\overline{x}_0$ such that the solution of (2.1.93) with starting value $\overline{x}_0$ and y_0 fixed tends to the equilibrium x_0 being stable. When the reduced system (2.1.89) with initial values (x_0, y_0) remains outside an ϵ-independent neighbourhood of a point with $\det(A) = 0$ for a time interval $[t_0, t_2]$, then for a time interval $[t_1, t_2]$ with $t_1 - t_0 > 0$ and independent of ϵ

$$|x^{(1)}(t) - x^{(2)}(t)| = O(\epsilon) \quad \text{and} \quad |y^{(1)}(t) - y^{(2)}(t)| = O(\epsilon).$$

Behaviour in a neighbourhood of singular points. Let us assume that the solution of the reduced system (2.1.89) passes a singular point $p \in \mathcal{F}$ where $\det(A) = 0$. By bringing F and G into a Taylor expansion at $(x,y) = p$ and making a linear transformation we obtain the following system of equations

$$\epsilon \frac{d\xi_1}{dt} = \xi_1^2 + \eta_1 + \sum_{j=2}^{n} b_{1j}\eta_j + \sum_{j=1}^{n} c_{1j}\xi_1\eta_j + \sum_{i=2}^{m} e_{1i}\xi_1\xi_i + d_{11}\xi_1^3 + \cdots,$$

$$\epsilon\frac{d\xi_i}{dt} = \sum_{j=2}^{m} a_{ij}\xi_j + \sum_{j=1}^{n} b_{ij}\eta_j + \sum_{j=1}^{n} c_{ij}\xi_1\eta_j + \sum_{j=1}^{m} e_{ij}\xi_1\xi_j + d_{i1}\xi_1^3 + \cdots,$$

$$i = 2,\ldots,m,$$

$$\frac{d\eta_1}{dt} = 1+\alpha_{11}\xi_1+\cdots, \tag{2.1.97}$$

$$\frac{d\eta_j}{dt} = \alpha_{j1}\xi_1+\cdots, \quad j = 2,\ldots,n.$$

Taking ξ_1 as independent variable, we have the $(m+n-1)$-dimensional system

$$\frac{d\xi_i}{d\xi_1} = \frac{\sum_{j=2}^{m} a_{1j}\xi_j+\cdots}{\xi_1^2+\eta_1+\cdots}, \quad i = 2,\ldots,m, \tag{2.1.98a}$$

$$\frac{d\eta_j}{d\xi_1} = \epsilon\frac{\delta_{1j}+\alpha_{1j}+\cdots}{\xi_1^2+\eta+\cdots}, \quad j = 1,\ldots,n, \tag{2.1.98b}$$

where δ_{ij} denotes the Kronecker δ.

We analyse trajectories of (2.1.98) for the interval $\xi_1 \in (-\rho,\rho)$, where ρ is a sufficiently small positive number independent of ϵ. This interval is divided into three parts:

$$(-\rho,-\sigma_1(\epsilon)),\ (-\sigma_1(\epsilon),\sigma_2(\epsilon)) \quad \text{and} \quad (\sigma_2(\epsilon),\rho), \tag{2.1.99}$$

where the positive order functions $\sigma_i(\epsilon)$ are $o(1)$. Later on a specific choice will be made.

The interval $(-\rho,-\sigma_1(\epsilon))$. It is supposed that

$$\xi_i = \xi_i^{(0)}(\xi_1)+\epsilon\xi_i^{(1)}(\xi_1)+\epsilon^2\xi_i^{(2)}(\xi_1)+S_{i1}(\xi_1;\epsilon), \quad i = 2,\ldots,m, \tag{2.1.100a}$$

$$\eta_j = \eta_j^{(0)}(\xi_1)+\epsilon\eta_j^{(1)}(\xi_1)+\epsilon^2\eta_j^{(2)}(\xi_1)+R_{j1}(\xi_1;\epsilon), j = 1,\ldots,n. \tag{2.1.100b}$$

The functions $\xi_i^{(0)}$ and $\eta_j^{(0)}$ satisfy (2.1.97) with $\epsilon = 0$. The higher order terms $\xi_i^{(1)}$, $\xi_i^{(2)}$, $\eta_j^{(1)}$ and $\eta_j^{(2)}$ follow from (2.1.98):

$$\xi_i(\xi_1) = A_{i2}\xi_1^2+A_{i3}\xi_1^3+\cdots+\epsilon(B_i\xi_1^{-1}+C_i \ln|\xi_1|)+ \\ +\epsilon^2 D_i\xi_1^{-4}+S_{i1}(\xi_1;\epsilon), \tag{2.1.101a}$$

$$\eta_1(\xi_1) = -\xi_1^2+\left(\frac{2}{3}\sum_{j=2}^{n}\alpha_{j1}b_{1j}+c_{11}-d_{11}-\sum_{j=2}^{m}e_{1j}A_{j2}\right)\xi_1^3+\cdots \\ +\epsilon\left(-\frac{1}{2}\xi_1^{-1}-\frac{1}{2}\sum_{j=2}^{n}\alpha_{j1}b_{1j}\ln|\xi_1|\right)-\frac{1}{8}\epsilon^2\xi_1^4 \\ +R_{11}(\xi_1,\epsilon), \tag{2.1.101b}$$

$$\eta_j(\xi_1) = -\frac{2}{3}\alpha_{j1}\xi_1^3+\cdots+\frac{1}{2}\epsilon\alpha_{j1}\ln|\xi_1|+R_{j1}(\xi_1;\epsilon). \tag{2.1.101c}$$

The interval $(-\sigma_1(\epsilon),\sigma_2(\epsilon))$. For this interval we introduce the local variables u_i, v_j and τ:

$$\xi_1 = \mu u_1, \quad \xi_i = \mu^2 u_i, \quad t = \mu^3\tau, \tag{2.1.102a}$$

$$\eta_1 = \mu^2 v_1, \quad \eta_j = \mu^3 v_j, \quad \mu^3 = \epsilon, \tag{2.1.102b}$$

so that (2.1.97) transforms into

$$\frac{du_1}{d\tau} = u_1^2 + v_1 + \mu + \\ +(\sum_{j=2}^{n} b_{1j}v_j + c_{11}u_1v_1 + d_{11}u_1^3 + \sum_{j=2}^{m} e_{1j}u_1u_j) + \cdots, \tag{2.1.103a}$$

$$\mu\frac{du_i}{d\tau} = \sum_{j=2}^{m} a_{ij}u_j + b_{i1}v_1 + c_{i0}u_1^2 + \\ +\mu\sum_{j=2}^{n} b_{ij}v_j + c_{i1}u_1v_1 + d_{i1}u_1^3 + \sum_{j=2}^{m} e_{ij}u_1u_j) + \cdots, \tag{2.1.103b}$$

$$\frac{dv_1}{d\tau} = 1 + \mu\alpha_{11}u_1 + \cdots, \tag{2.1.103c}$$

$$\frac{dv_j}{d\tau} = \alpha_{j1}u_1 + \cdots. \tag{2.1.103d}$$

Likewise the system (2.1.98) can be transformed. Setting $\mu = 0$ we obtain the reduced system

$$\sum_{j=2}^{m} a_{ij}u_j + b_{i1}v_1 + c_{i0}u_1^2 = 0, \tag{2.1.104a}$$

$$\frac{dv_1}{du_1} = \frac{1}{u_1^2 + v_1}, \tag{2.1.104b}$$

$$\frac{dv_j}{du_1} = \frac{\alpha_{j1}u_1}{u_1^2 + v_1}. \tag{2.1.104c}$$

Writing the solutions as

$$v_1(u_1) = v_1^{(0)}(u_1) + \mu v_1^{(1)}(u_1) + r_1(u_1;\mu), \tag{2.1.105a}$$

$$v_j(u_1) = v_j^{(0)}(u_1) + r_j(u_1;\mu), \tag{2.1.105b}$$

$$u_i(u_1) = u_i^{(0)}(u_1) + \mu u_i^{(1)}(u_1) + s_i(u_1;\mu), \tag{2.1.105c}$$

we calculate $v_j^{(0)}$ and $u_i^{(0)}$ from (2.1.104) and next $v_j^{(1)}$ and $u_i^{(1)}$ from (2.1.103). Those particular solutions are selected that match the corresponding solutions for the interval $(-\rho,-\sigma_1(\epsilon))$. For these solutions asymptotic expressions for large positive and negative arguments are derived:

$$v_1^{(0)}(u_1)^- = u_1^2 - \frac{1}{2}u_1^{-1} - 8u_1^{-4} + O(u_1^{-7}), \tag{2.1.106a}$$

$$v_1^{(0)}(u_1)^+ = \alpha - u_1^{-1} + O(u_1^{-3}), \quad \alpha = 2.338..., \tag{2.1.106b}$$

$$u_i^{(0)}(u_1)^- = A_{i2}u_1^2 + B_i u_1^{-1} + D_i u_1^{-4} + O(u_1^{-7}), \tag{2.1.106c}$$

$$u_i^{(0)}(u_1)^+ = E_{i0}u_1^2 + O(1), \tag{2.1.106d}$$

$$v_1^{(1)}(u_1)^- = (\frac{2}{3}\sum_{j=2}^{n}\alpha_{j1}b_{1j} + c_{11} - \sum_{j=2}^{m} e_{1j}A_{j2})u_1^3 + \\ -\frac{1}{2}\sum_{j=2}^{n}\alpha_{j1}b_{1j}(\ln|u_1| + \ln\mu) + O(1), \tag{2.1.106e}$$

$$v_1^{(1)}(u_1)^+ = \alpha_{11} - d_{11} + \\ -\sum_{j=2}^{n} e_{1j}E_{j0}\ln|u_1| + \frac{1}{2}\alpha_{j1}b_{1j}\ln\mu + O(1), \tag{2.1.106f}$$

$$v_j^{(0)}(u_1)^- = -\frac{2}{3}\alpha_{j1}u_1^3 + \frac{1}{2}\alpha_{j1}\ln|u_1| + \frac{1}{2}\alpha_{j1}\ln\mu + \cdots, \tag{2.1.106g}$$

$$v_j^{(0)}(u_1)^+ = \alpha_{j1}\ln|u_1| + \frac{1}{2}\alpha_{j1}\ln\mu + \cdots. \tag{2.1.106h}$$

The interval $(\sigma_2(\epsilon),\rho)$. A first approximation is found from (2.1.98) with $\epsilon = 0$. This yields

$$\eta_j = 0, \quad j = 2,...,n, \tag{2.1.107a}$$

$$\xi_i = E_{i0}\xi_1^2 + \cdots, \quad i = 2,...,m, \tag{2.1.107b}$$

The higher order terms that match the solution for the interval $(-\sigma_1(\epsilon),\sigma_2(\epsilon))$ are

$$\xi_i(\xi_1;\epsilon) = E_{i0}\xi_1^2 + \cdots + S_{i2}(\xi_1;\epsilon), \quad i = 2,...,m, \tag{2.1.108a}$$

$$\eta_1(\xi_1;\epsilon) = \alpha\epsilon^{\frac{2}{3}} \frac{1}{3}(\alpha_{11} - d^1 - \sum_{j=2}^{m} e_{1j}E_{j0} + \frac{1}{2}\sum_{j=2}^{n} b_{1j}\alpha_{j1})\epsilon\ln\epsilon + \\ + \{-\xi_1^{-1} + (\alpha_{11} - d_{11} - \sum_{j=2}^{m} e_{1j}E_{j0})\ln\xi_1\}\epsilon + \\ + R_{12}(\xi_1;\epsilon), \tag{2.1.108b}$$

$$\eta_j(\xi_1;\epsilon) = -\frac{1}{6}\alpha_{j1}\epsilon\ln\epsilon + \alpha_{j1}\epsilon\ln\xi_1 + R_{j2}(\xi_1;\epsilon). \tag{2.1.108c}$$

We define a vector D^p belonging to the singular point p as follows

$$D_1^p = \alpha\epsilon^{\frac{2}{3}} - \frac{1}{3}(\alpha_{11} - d_{11} - \sum_{j=2}^{m} e_{1j}E_{j0} + \frac{1}{2}\sum_{j=2}^{n} b_{1j}\alpha_{j1})\epsilon\ln\epsilon, \tag{2.1.109a}$$

$$D_j^p = -\frac{1}{6}\alpha_{j1}\epsilon\ln\epsilon, \quad j = 2,...,n. \tag{2.1.109b}$$

The vector D^p is the displacement in η from $\xi_1 = -\sigma_1(\epsilon)$ to $\xi_1 = \sigma_2(\epsilon)$.

If one chooses

$$\sigma_1(\epsilon) = \epsilon^{2/7} \text{ and } \sigma_2(\epsilon) = \epsilon^{2/9},$$

then all remainder terms in the three intervals are $O(\epsilon)$. More specifically it concerns the terms

$$R_{j1}, S_{i1},\ i = 2,...,m,\ j = 1,...,n \text{ for } \xi_1 \in (-\rho, -\sigma_1(\epsilon)),$$

$$\mu^2 r_1, \mu^3 r_j, \mu^2 s_i,\ i = 2,...,m,\ j = 2,...,n \text{ for } \xi_1 \in (-\sigma_1(\epsilon), \sigma_2(\epsilon)),$$

$$R_{j2}, S_{i2},\ i = 2,...,m,\ j = 1,...,n \text{ for } \xi_1 \in (\sigma_2(\epsilon), \rho).$$

Pontryagin gives formula's for D^p in terms of the coefficients of the Taylor expansions for F and G at p. He also proves that the expansions (2.1.100), (2.1.105) and (2.1.108) converge. The linear coordinate transformation $(x,y) \rightarrow (\xi,\eta)$ did not mix the "fast" and "slow" subspaces. Thus, the displacement vector D^p gives the correction to the position of the "fast" subspace, which in the discontinuous approximation goes exactly through p and satisfies $y = y_p$.

The Poincaré mapping. Let us take the $(n-1)$-dimensional transversal intersection Σ in $\mathcal{F}$ with $\det(A) \neq 0$ for all $s \in \Sigma$. The discontinuous limit trajectories generate a mapping

$$P: \Sigma \rightarrow \Sigma.$$

This mapping or a finite repetition of it may have a fixed point that corresponds with a discontinuous periodic solution L_0. When this fixed point is stable, existence of a periodic solution of the complete system with $0 < \epsilon << 1$ can be proved, as we will see. It is remarked that the mapping may generate chaotic solutions. In fig. 2.1.7 we sketch the case that Σ is a 2-dimensional manifold ($n=3$) with a fixed point $\bar{s} \in \Sigma$ of saddle point type. For a point $s_0 \in \Sigma$ at the stable manifold the iterates satisfy $s_k \rightarrow \bar{s}$ for $k \rightarrow \infty$. For a point $s_0 \in \Sigma$ at the unstable manifold, the iterates will leave a neighbourhood of $\bar{s}$. The stable and unstable manifolds are intertwined in a complicated way, such that the two manifolds intersect. Taking a starting point $s_0 \in \Sigma$ at this intersection, we conclude that also the next point $s_1 = Ps_0$ is at both manifolds. Consequently, the unstable manifold must return to a neighbourhood of the fixed point intersecting infinitely many times the stable manifold, see fig. 2.1.7. The same argument holds for the stable manifold, as one considers the iterates s_k backward for $k \rightarrow -\infty$.

As seen from fig. 2.1.8 the rectangle R is compressed in one direction and expanded in the other direction under the iterations of the mapping P. After a given number k of iterations $P^k R$ and R overlap and have two distinct rectangles a and b in common. The mapping P^k is a so-called horse-shoe map, see appendix C. For any bi-infinite string of a's and b's, e.g.

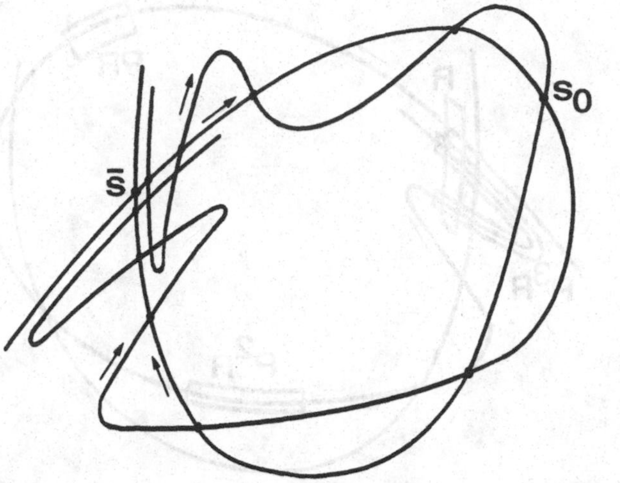

Fig.2.1.7 Paths of the iterations at the unstable and unstable manifolds. The assumption that s_k is at both manifolds implies that s_{k+1} and s_{k-1} are also at both manifolds.

...abbaaabaabbab...,

we can find a point in $a \cup b$ such that the sequence of points, formed by the forward and backward interates of P^k, are in a or b in the same order. This correspondence between trajectories of the system and the string of symbols "a" and "b" is a useful tool in the analysis of chaotic dynamics. We will deal with this subject in more detail in chapter 4. In section 2.6.1 we analyse chaotic dynamics of a system with Σ being a curve ($n = 2$).

Existence of a periodic solution. Let us assume that the reduced system (2.1.89) has a discontinuous periodic solution denoted by L_0 in the phase space $\mathbb{R}^{m+n}$. As an example we take the case where L_0 contains two leaving points p_1 and p_2. The transition from the fast action to the slow action is at r_1 and r_2.

We define a set of points V_i in a neighbourhood of r_i as follows. Let $U_i \subset \mathcal{F}$ be the set of points in a neighbourhood of p_i satisfying $\det(A) = 0$, then

$$V_i = \{s = (x_s, y_s) | F(s) = 0,\ (x, y_s) \in U_i \text{ for some } x\}. \qquad (2.1.110)$$

At the next point we deviate from the method of Mishchenko and Rosov. We define the mappings

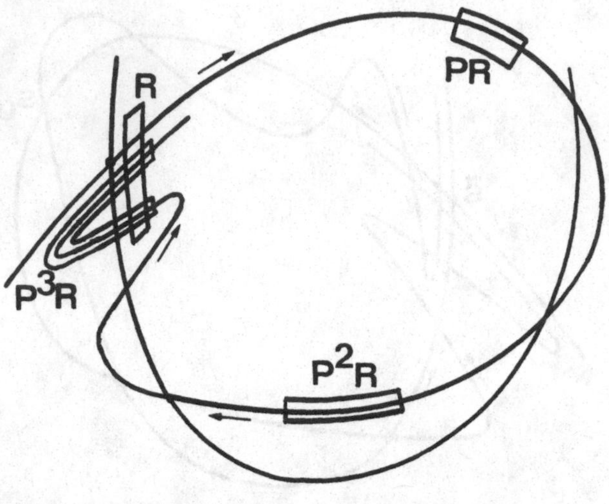

Fig.2.1.8 The horse shoe map. The rectangle R is stretched along the stable manifold and compressed perpendicular to it. After three mappings the image P^3R intersects R.

$$N_1: V_2 \rightarrow U_1, \tag{2.1.111a}$$

$$N_2: V_1 \rightarrow U_2 \tag{2.1.111b}$$

by the trajectories of the reduced system (2.1.89) which starting at $s \in V_2$ arrive in a point $s \in U_1$, see fig. 2.1.9. In a similar way N_2 is defined, Let $\mathcal{U}_i$ and $\mathcal{V}_i$ be the $(n-1)$-dimensional hyperplanes tangential to U_i and V_i in p_i and r_i. Let M_i be the linearization of N_i in these spaces. We introduce a projection into the Y-space. For $\mathcal{U}_i$ and $\mathcal{V}_i$ this yields after translation of p_i and r_i to the origin the subspaces with $\mathcal{U}'_i = \mathcal{V}_i$. Moreover, we have the induced mapping M_i^y. We define

$$\pi_1 = M_1^y M_2^y \text{ and } \pi_2 = M_2^y M_1^y. \tag{2.1.112}$$

These linear mappings have the origin as only fixed point (being the orbit L_0).

Let a trajectory of (2.1.89) through a point $s \in \mathcal{F}$ near r_i intersect V_i in w without leaving a neighbourhood of r_i. Then we consider the mapping

$$K_i: s \mapsto w. \tag{2.1.113}$$

Carrying out the procedure of linearization, projection and translation we obtain

$$K_i^{\epsilon}: Y \to \mathcal{V}_i \; (= \mathcal{U}'_i). \tag{2.1.114}$$

Finally, we introduce the nonlinear mappings $\pi_i^{\epsilon}: Y \to \mathcal{U}'_i$ as

$$\pi_1^{\epsilon} y = M_1^{\epsilon} K_2^{\epsilon} [M_2^{\epsilon} K_1^{\epsilon} (y + D^{P_1} p) + D^{P_2}] \tag{2.1.115a}$$

$$\pi_2^{\epsilon} y = M_2^{\epsilon} K_1^{\epsilon} [M_1^{\epsilon} K_2^{\epsilon} (y + D^{P_2}) + D^{P_1}] \tag{2.1.115b}$$

with D^{P_i} given by (2.1.109).
The relation between trajectories of (2.1.88) near L_0 and these mappings is given in the following theorem.

Theorem 2.1.6. Let $y_1 \in \mathcal{U}'_1$ be a fixed point of π_1^{ϵ} and s_1 its corresponding point in $\mathcal{U}_1$. Let $\{x^{(0)}(t), y^{(0)}(t)\}$ be a trajectory of (2.1.89) satisfying $\{x^{(0)}(t_{s_1}), y^{(0)}(t_{s_1})\} = s_1$. Then (2.1.88) has a closed trajectory L_ϵ satisfying

$$|x(t;\epsilon) - x^{(0)}(t)| = O(\epsilon), \quad |y(t;\epsilon) - y^{(0)}(t)| = O(\epsilon) \tag{2.1.116}$$

for a time interval $[t_0, t_{s_1} - \delta]$ with $\delta > 0$ and independent of ϵ. Moreover, L_ϵ tends to L_0 as $\epsilon \to 0$.

For the proof we refer to Mishchenko and Rosov (1980). It is remarked that π_1^{ϵ} must have a fixed point as it maps a compact domain of Y into itself.

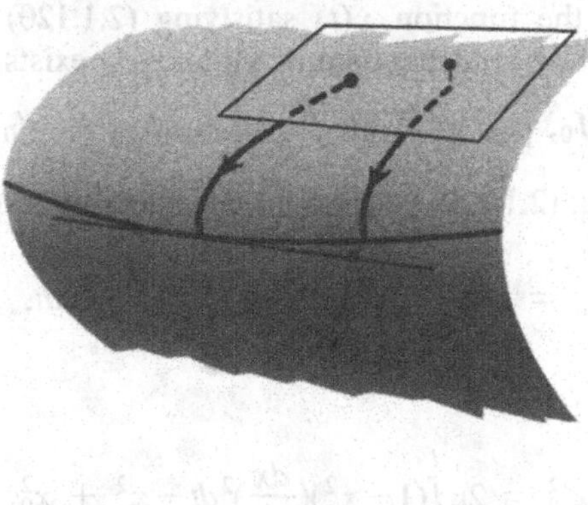

Fig.2.1.9 The flow from V_1 to U_2 restricted to the manifold $\mathcal{F}$ and the mapping induced by this flow in the tangential spaces $\mathcal{V}_1$ and $\mathcal{U}_2$.

Computation of the period $T(\epsilon)$. The time T_0 spent outside a $o(1)$-neighbourhood of the singular points is computed in the way we described in section 1.4. The time $T^{(\rho)}$ needed to pass the interval $-\rho \leqslant \xi_1 \leqslant \rho$ has to be calculated separately from the local expansions for the three subintervals:

$$T^{(\rho)} = \int_{-\rho}^{\rho} \frac{d\eta_1/d\xi_1}{d\eta_1/dt} d\xi_1. \tag{2.1.117}$$

This expression contains a contribution from the regular expansion, we already accounted for in T_0, so

$$T^{(\rho)} = T_0^{(\rho)} + T_p^{(\rho)}(\epsilon) \text{ with } T_p^{(\rho)}(\epsilon) = O(\epsilon^{2/3}) \tag{2.1.118}$$

and

$$T(\epsilon) = T_0 + T_{p_1}^{(\rho)}(\epsilon) + T_{p_2}^{(\rho)}(\epsilon) + O(\epsilon). \tag{2.1.119}$$

Explicit expressions for $T_{p_i}^{(\rho)}(\epsilon)$ are given in Mishchenko and Rosov (1980).

2.1.5 The analytical method of Cartwright

Cartwright (1952) analyses the Van der Pol equation

$$\frac{d^2x}{dt^2} + \mu(x^2-1)\frac{dx}{dt} + x = 0, \quad \mu >> 1 \tag{2.1.120}$$

in the x,t-plane. Her method is based on locally valid estimates of x and its derivatives, which are obtained from the following theorem proved in Cartwright (1952):

Theorem 2.1.7 For the function $x(t)$ satisfying (2.1.120) with $x(0)=x_0$ and $x'(0)=x'_0$ a constant M_0 independent of x_0 and x'_0 exists such that

$$|x(t)| < M_0, \ |x'(t)| < M_0\mu \ \text{ for } t > t_0(M_0, x_0, x'_0).$$

In addition to Eq. (2.1.120) Cartwright considers the integrated equation

$$x'(t) - x'_0 = \mu(x - \tfrac{1}{3}x^3 - x_0 + \tfrac{1}{3}x_0^3) - \int_{t_0}^{t} x\,dt, \tag{2.1.121}$$

the energy equation

$$\left(\frac{dx}{dt}\right)^2 - x'^2_0 = 2\mu\int_{t_0}^{t}(1-x^2)\left(\frac{dx}{dt}\right)^2 dt - x^2 + x_0^2, \tag{2.1.122}$$

as well as the differentiated equation

$$\frac{d^3x}{dt^3} + \mu(x^2-1)\frac{d^2x}{dt^2} + 2\mu x\left(\frac{dx}{dt}\right)^2 + \frac{dx}{dt} = 0. \tag{2.1.123}$$

The value of $x(t)$ and its derivative in A is denoted by a, a', etc., see fig. 2.1.10. Its is noted that

$h'=0$, $c=1$ and $a=0$.

In a sequence of 12 lemma's x, x', x'' and t are estimated in the different intervals. Roughly the line of argument is as follows. In lemma 1 the energy equation is used to prove that a solution, starting in the maximum H not to far above $x=1$, arrives in C with c' bounded by a number that only depends on h. In lemma 2 lower bounds for $x'(t)$ and t are given for the interval HC. Lemma's 3, 4 and 5 show that for the intervals HY, YZ and ZE the lower bounds given by lemma 2 are indeed good approximations of $x'(t)$ and t for $h>3/2$ and $x(t)$ not too close to 1. Lemma 6 and 7 deal, respectively, with the arcs EC and CF just before and after C. They prove that the time needed to pass the strip $|x-1|<\Delta\mu^{-2/3}$ is at most $M\Delta^2\mu^{-1/3}$ for every $\Delta>\Delta_0$ and $\mu\geqslant\mu_0(\Delta)$ with M fixed. Lemma 8 gives an estimate of the time needed to reach A' from C.

As an illustration of the method we state lemma 9 and also give the proof.

Lemma 2.1.1 A solution of (2.1.120), that starts in A with $a'>3/\mu$ reaches B with

$$|b'-a'-\frac{2}{3}\mu|\leqslant t_{ab}<M\frac{\ln\mu}{\mu}. \tag{2.1.124}$$

Proof Let G be the point with $g=1/2$. Then for the interval AG we have because of (2.1.120) that

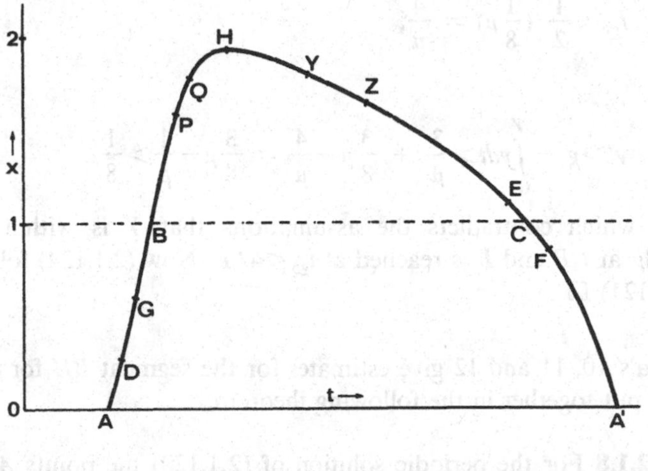

Fig.2.1.10 The solution of the Van der Pol equation and the successive points between which different types of estimates are made in the method of Cartwright.

$$\frac{d^2x}{dt^2} \geqslant \frac{3}{4}\mu\frac{dx}{dt} - x$$

and so $x'(t)$ increases provided that $x<3/4\mu a'$. Since x is less than $1/2$ at AG and $3/4\mu a'>9/4$, this holds for the complete segment AG, so that

$$t_{ag} < \frac{1}{2a'} \leqslant \frac{1}{6}\mu .$$

Because of (2.1.121) the following inequality also holds at AG

$$x'(t) > \frac{3}{\mu} + \mu x(1 - \frac{1}{3}x^2) - \int_a^x dt >$$

$$\frac{3}{\mu} + \frac{11}{12}\mu x - \frac{1}{6}\mu x = \frac{3}{\mu} + \frac{3}{4}\mu x.$$

Consequently, we have that

$$g' > \frac{3}{\mu} + \frac{3}{8}\mu$$

and

$$t_{ag} = \int_a^g \frac{1}{dx/dt}dx < \int_0^{1/2} \frac{dx}{\frac{3}{\mu} + \frac{3}{4}\mu x} < \frac{M\ln\mu}{\mu} \quad \text{for } \mu>\mu_0.$$

Furthermore, it is seen that $x'(t)>1/8\mu$ near G. If ever $x'(t)=1/8\mu$ at GB for some point Y, then for the first point with this property

$$t_{gy} < \frac{1}{2}/(\frac{1}{8}\mu) = \frac{4}{\mu},$$

and

$$y' > g' - \int_G^Y y dt > \frac{3}{\mu} + \frac{3}{8}\mu - \frac{4}{\mu} = \frac{3}{8}\mu - \frac{1}{\mu} > \frac{1}{8}\mu$$

for $\mu>2$, which contradicts the assumptions that Y is within GB. Thus, $x'(t)>1/8\mu$ at GB and B is reached at $t_{gb}<4/\mu$. Now (2.1.124) follows directly from (2.1.121) □

Lemma's 10, 11 and 12 give estimates for the segment BH for b' large. The pieces are put together in the following theorem.

Theorem 2.1.8 For the periodic solution of (2.1.120) the points A, B, C and H and the period T satisfy

$$\frac{M}{\Delta\mu^{1/3}} \leqslant |c'| \leqslant \frac{M\Delta^{1/2}}{\mu^{1/3}}, \quad |a' - \frac{2}{3}\mu| \leqslant M\Delta^{1/2}\mu^{-1/3},$$

$$|b'-\frac{4}{3}\mu|\leqslant M\Delta^{1/2}\mu^{-1/3},\quad |h-2|\leqslant M\Delta^{1/2}\mu^{-4/3}$$

and

$$|T-\mu(\frac{3}{2}-\ln 2)|<M\Delta^2\mu^{-1/3}$$

for

$$\Delta>\Delta_0 \text{ and } \mu>\mu_0(\Delta).$$

Later on in the article, Cartwright (1952) shows that Δ can be replaced by α with $-\alpha$ being the first zero of the Airy function.

Exercises

2.1.1 Consider the system (2.1.7) with

$$F(x) = x^5 + \alpha x^3 + \beta x.$$

For which values of α and β does the system have one or more periodic solutions being relaxation oscillations?

2.1.2 Check the inner product (n,v) and the direction of the normal vector in La Salle's method for the segment B_1C_1.

2.1.3 Derive asymptotic expressions for amplitude and period of the Van der Pol relaxation oscillator from the formula's of Ponzo and Wax.

2.1.4 Sketch a closed trajectory of (2.1.88) for $m=1, n=2$ with two jumps and follow the subsequent steps in the mapping π_1^ϵ of (2.1.115a).

2.2 ASYMPTOTIC SOLUTION OF THE VAN DER POL EQUATION

Using the method of matched asymptotic expansions, see section 1.5, one may solve the Van der Pol equation (1.1.3) for $\nu>>1$ in three different ways. In section 2.2.1 the asymptotic solution in the x,t-plane is constructed. It gives a good introduction to the problem of the Van der Pol oscillator with a sinusoidal forcing term. Dorodnicyn (1947) approaches trajectories in the $x,\dot{x}$-plane asymptotically by local solutions in four regions. In section 2.2.2 we carry out this method and modify it by introducing an additional region in order to have a correct asymptotic matching. In section 2.2.3 we present the method of Waldvogel (1985) yielding local asymptotic solutions in the Lienard plane. Crucial for the three methods is the correct Ansatz. That is to start with the appropriate local solution and to expand formally the appropriate functions and unknown constants. There is no rule for this and, as O'Malley (1982) states, it has aspects which make it hard to deal with it systematically. For this reason we work out the three methods in detail. Finally, in section 2.2.4 we discuss methods to approximate period and amplitude of the Van der Pol relaxation oscillator and refer to numerical and asymptotic results.

2.2.1. The physical plane

Using singular perturbation methods Carrier and Lewis (1953) and Kevorkian and Cole (1981) construct an asymptotic approximation in the physical plane (x,t-plane) of the periodic solution of the Van der Pol equation

$$\epsilon \frac{d^2 x}{dt^2} + (x^2-1)\frac{dx}{dt} + x = 0, \quad 0<\epsilon<<1, \tag{2.2.1}$$

see fig.2.2.1. MacGillivray (1983) proves the correctness of the boundary layer approximation.

We work as follows. First an approximation is made in a time interval, where $x'(t)=O(1)$. At the point t_0, where this approximation breaks down, because of the fact that $x'(t)\to\infty$ as $(t,\epsilon)\to(t_0,0)$, a new local approximation is introduced and it is matched with the previous one. This procedure is continued in the appropriate time scales until the solution returns in a region where the first approximation holds.

For $\epsilon=0$ we have the following reduced equation

$$(x_0^2-1)\frac{dx_0}{dt} + x_0 = 0 \tag{2.2.2a}$$

or

$$t = \ln|x_0| - \frac{1}{2}(x_0^2-1) + A. \tag{2.2.2b}$$

For $t<0$ we choose the branch $x_0>0$ with integration constant $A=0$. For $t\uparrow 0$ we have asympotically

$$x_0 \approx 1 + \sqrt{-2t},$$

so that it is incorrect to neglect the second derivative of (2.2.1) near $(x,t)=(1,0)$. In a local analysis of the asymptotic behaviour of the solution one applies a stretching transformation

$$t = \xi\epsilon^{2/3}, \quad x = 1 + \epsilon^{1/3} v(\xi). \tag{2.2.3}$$

The powers of ϵ are chosen such that formally the three terms of (2.2.1) are of the same order of magnitude. After substitution of (2.2.3) the leading terms of (2.2.1) form the equation

$$\frac{d^2 v_0}{d\xi^2} + 2v_0\frac{dv_0}{d\xi} + 1 = 0 \quad \text{or} \quad \frac{dv_0}{d\xi} + v_0^2 + \xi = B. \tag{2.2.4a,b}$$

Looking for a solution $v_0 = z'(\xi)/z(\xi)$ we find that $z(\xi)$ has to satisfy the Airy

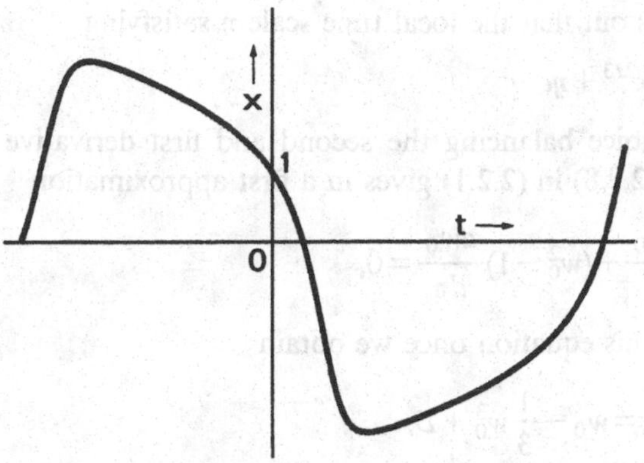

Fig.2.2.1 The periodic solution in the physical plane. Local approximations are made for the interval where the solution decreases from the value 2 to 1, the interval near $t=0$ where the solution remains near the value 1 and the boundary layer interval where it rapidly decreases from 1 to -2.

equation

$$\frac{d^2z}{d\xi^2}-(B-\xi)z=0 \tag{2.2.5}$$

for arbitrary B. Its general solution reads

$$z(\xi)=\lambda\{Ai(B-\xi)+CBi(B-\xi)\} \tag{2.2.6}$$

with λ and C arbitrarily chosen, $Ai(s)$ and $Bi(s)$ denote the Airy functions, see Appendix A. For $\xi\rightarrow-\infty$ the local asymptotic solution has to match the outer solution $x_0(t)$ away from $t=0$. This is the case for

$$v_0(\xi)\approx\sqrt{-\xi} \text{ with } \xi<<-1.$$

The expression (2.2.6) has this asymptotic behaviour for $B=C=0$, so that

$$v_0(\xi)=-Ai'(-\xi)/Ai(-\xi). \tag{2.2.7}$$

The Airy function $Ai(z)$ has its fist zero for $z=-\alpha=-2.33811$. As ξ approaches this value from below we have that

$$v_0(\xi)\approx(\xi-\alpha)^{-1}.$$

This changes the balance in the order of magnitudes of d^2x/dt^2, dx/dt, x and x^2-1. Therefore, we have again to find an appropriate local coordinate at

$t = \alpha\epsilon^{2/3}$. It turns out that the local time scale η satisfying

$$t = \alpha\epsilon^{2/3} + \eta\epsilon \tag{2.2.8}$$

is the correct choice balancing the second and first derivative of x in (2.2.1). Substutition of (2.2.8) in (2.2.1) gives in a first approximation

$$\frac{d^2 w_0}{d\eta^2} + (w_0^2 - 1)\frac{dw_0}{d\eta} = 0, \tag{2.2.9}$$

and integrating this equation once we obtain

$$\frac{dw_0}{d\eta} = w_0 - \frac{1}{3}w_0^3 + D. \tag{2.2.10}$$

For $\eta \to -\infty$ this equation has to take the form (2.2.4b). Setting

$$w_0 = 1 + \epsilon^{1/3} v_0(\eta)$$

we find

$$\epsilon^{1/3}\frac{dv_0}{d\eta} - \frac{2}{3} + \epsilon^{2/3} v_0^2 - D = O(\epsilon) \tag{2.2.11}$$

and, as $\xi = \alpha + \eta\epsilon^{1/3}$, D must take the value $-2/3 - \alpha\epsilon^{2/3}$. Eq.(2.2.10) has for this value of D the following solution

$$E - \eta = \frac{1}{1 - w_0} + \frac{1}{3}\ln\left\{\frac{w_0 + 2 + 1/3\alpha\epsilon^{2/3}}{1 - w_0}\right\}. \tag{2.2.12}$$

For $\eta \to -\infty$ w_0 has the correct matching behaviour, if one chooses $E = 0$. On the other hand for $\eta \to \infty$ we have

$$w_0 = -2 - \frac{1}{3}\alpha\epsilon^{2/3} + O(\exp(-3\eta)). \tag{2.2.13}$$

Then the solution approaches a region, where it is approximated by the regular degenerate equation (2.2.2). Substitution of

$$(x,t) = (-2 - 1/3\alpha\epsilon^{2/3}, \alpha\epsilon^{2/3})$$

yields

$$A = \frac{3}{2} - \ln 2 + \frac{3}{2}\alpha\epsilon^{2/3}. \tag{2.2.14}$$

At $(x,t) = (-1, 1/2T)$ the solution has completed half its period. Using (2.2.2b) and (2.2.14) we find

$$T \approx 3 - 2\ln 2 + 3\alpha\epsilon^{2/3}. \tag{2.2.15}$$

It is noted that compared with the original method by Carrier and Lewis we deleted a local asymptotic solution valid in an ϵ-neighbourhood of

$$(x,t) = (-2 - \frac{1}{3}\alpha\epsilon^{2/3}, \alpha\epsilon^{2/3} + \frac{1}{3}\epsilon\ln\epsilon),$$

where the boundary layer solution $w_0(\eta)$ and the outer solution $x_0(t)$ match.

2.2.2 The phase plane

Dorodnicyn (1947) takes Eq.(2.1.1) as the starting point of his asymptotic analysis of the Van der Pol equation. Since

$$\frac{d^2x}{d\tau^2}=\frac{dp}{d\tau}=\frac{dp}{dx}\frac{dx}{d\tau}=p\frac{dp}{dx}, \tag{2.2.16}$$

Eq.(2.2.1) is equivalent with

$$p\frac{dp}{dx}+\nu(x^2-1)p+x=0. \tag{2.2.17}$$

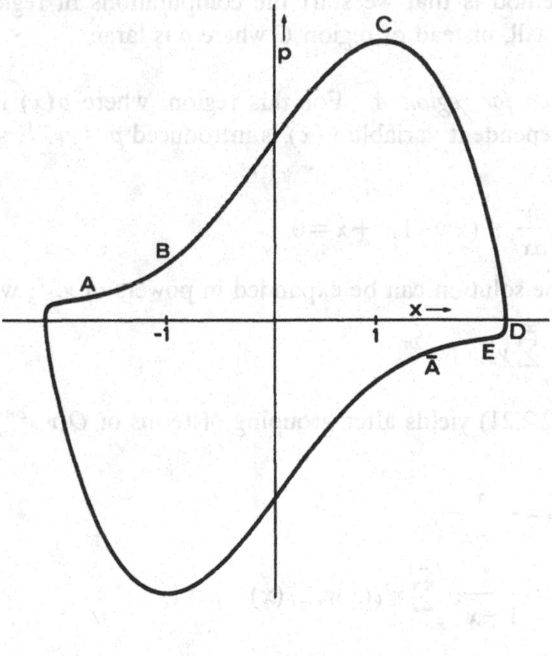

Fig.2.2.2 The closed trajectory in the phase plane. At each region a local asymptotic approximation is made. Intergration constants are determined by matching approximations of adjacent regions.

Van der Pol (1926) already noticed that the behaviour of the periodic solution for $\nu >> 1$ is described by the reduced equation

$$p\frac{dp}{dx} + \nu(x^2-1)p = 0 \tag{2.2.18}$$

in a region of the phase plane where $|p|$ is large, and by

$$\nu(x^2-1)p + x = 0 \tag{2.2.19}$$

in a region where $|p|$ and $|dp/dx|$ are small. The regions, where these approximate solutions are valid, do not overlap and it was, at that time, not clear how they could be matched. Dorodnicyn (1947) introduced two more regions, in which he gives different asymptotic solutions for (2.2.17). However, the problem of matching the four local solutions was still not solved in a satisfactory way. In this section, special attention will be given to the last point. It is possible to apply a well-founded method of matching, when a fifth region with its own local solution is added, see fig.2.2.2. Another important difference with Dorodnicyn's method is that we start the computations in region A, where p and dp/dx are small, instead of region C where p is large.

Asymptotic solution for region A. For this region, where $p(x)$ is of the order $O(\nu^{-1})$, a new dependent variable $y(x)$ is introduced $p = y\nu^{-1}$, so that (2.2.17) transforms into

$$\nu^{-2}y\frac{dy}{dx} + (x^2-1)y + x = 0. \tag{2.2.20}$$

Assuming that the solution can be expanded in powers of ν^{-2}, we write

$$y = \sum_{n=0}^{\infty} y_n(x)\nu^{-2n}. \tag{2.2.21}$$

Substitution in (2.2.21) yields after grouping of terms of $O(\nu^{-2m})$ the following recurrent system

$$y_0(x) = \frac{x}{1-x^2}, \tag{2.2.22a}$$

$$y_j(x) = \frac{1}{1-x^2}\sum_{k=0}^{j-1} y'_k(x)y_{j-k}(x) \quad j = 1,2,.... \tag{2.2.22b}$$

For $j = 1$ we have

$$y_1(x) = x(x^2+1)/(1-x^2)^4.$$

Asymptotic solution for region B. The expansion (2.2.21) breaks down as $x \uparrow -1$. In order to analyse the local behaviour we introduce the local variables u and Q by

$$x = -1 - u\,\nu^{\alpha}, \quad p = \nu^{\beta}Q(u;\nu), \quad \alpha,\beta < 0. \tag{2.2.23}$$

Eq.(2.2.17) then transforms into an equation for $Q(u)$ with the three terms of the order $O(1)$ for

$$2\beta-\alpha=1+\alpha+\beta=0.$$

Hence, the choice $(\alpha,\beta)=(-2/3,-1/3)$ will result in the appropiate local variables for the asymptotic solution in region B. This leads to a differential equation for $Q(u;\nu)$ with $\nu^{-2/3}$ as small parameter. We assume that Q can be expanded as

$$Q(u;\nu)=\sum_{j=0}^{\infty} Q_j(u)\nu^{-2j/3}. \tag{2.2.24}$$

Then the coefficients of this expansion satisfy the recurrent system

$$Q_0\frac{dQ_0}{du}-2uQ_0+1=0, \tag{2.2.25a}$$

$$Q_0^2\frac{dQ_1}{du}-Q_1=u^2Q_0^2-uQ_0, \tag{2.2.25b}$$

$$Q_0^2\frac{dQ_n}{du}-Q_n = u^2Q_0Q_{n-1}-\sum_{k=1}^{n-1}Q_0Q_k\frac{dQ_{n-k}}{du}. \tag{2.2.25c}$$

Eq.(2.2.25a) is transformed into a Riccati equation by substituting

$$Q_0(u)=u^2-z(u) \tag{2.2.26}$$

and by interchanging the roles of the independent variable u and the new dependent variable z:

$$\frac{du}{dz}-u^2+z=0. \tag{2.2.27}$$

This equation is in turn related to the Airy equation: substitution of $u=-V'(z)/V(z)$ yields

$$\frac{d^2V}{dz^2}-zV=0. \tag{2.2.28}$$

Thus, $z=z(u)$ is the inverse of

$$u=-\frac{C_0Ai'(z)+D_0B'_i(z)}{C_0Ai(z)+D_0Bi(z)}, \tag{2.2.29}$$

where $Ai(z)$ and $Bi(z)$ denote the Airy functions and C_0 and D_0 are arbitrary integration constants. The solution of (2.2.25b) takes the form

$$Q_1(u)=\frac{1}{A(u)}[C_1+\int_0^u A(v)\{v^2-\frac{v}{Q_0(v)}\}dv] \tag{2.2.30}$$

with

$$A(u)=\exp\{-\int_0^u\frac{dv}{Q_0^2(v)}\}. \tag{2.2.31}$$

Matching the asymptotic solutions for regions A and B. It is easily verified that the expansions (2.2.21) and (2.2.24) both converge asymptotically in an intermediate region with a local coordinate s satisfying

$$x=-1-s\delta(\nu) \tag{2.2.32}$$

with $\delta(\nu)$ such that

$$\delta(\nu) \quad \text{and} \quad \nu^{-2/3}/\delta(\nu)\to 0 \quad \text{as} \quad \nu\to\infty.$$

For the special choice $\delta(\nu)=\nu^{-1/3}$ expansion (2.2.21) transforms into

$$p=\frac{1}{2}s^{-1}\nu^{-2/3}+\frac{1}{4}\nu^{-1}-\frac{1}{8}s\nu^{-4/3}+O(\nu^{-5/3}). \tag{2.2.33}$$

The leading term or (2.2.24) has to be $O(\nu^{-1/3})$ after substitution of (2.2.32) with $\delta(\nu)=\nu^{-1/3}$. From (2.2.26) we see that necessarily $u\to\infty$ as $z\to\infty$. Using the theory of Airy functions it is concluded from (2.2.29) that D_0 has to vanish. In fact we have

$$u=-Ai'(z)/Ai(z)$$

and so

$$u=z^{1/2}\{1+\frac{1}{4}z^{-3/2}+O(z^{-3})\}, \quad z>>1. \tag{2.2.34}$$

Thus, substitution in (2.2.26) yields

$$\nu^{-1/3}Q_0(\nu^{1/3}s)=\frac{1}{2}s^{-1}\nu^{-2/3}-\frac{1}{8}s^{-4}\nu^{-5/3}+O(\nu^{-8/3}). \tag{2.2.35}$$

Regarding the order of magnitude of (2.2.30) for $u=\nu^{1/3}s$ we conclude that Q_1 matches (2.2.33) if

$$C_1=-\int_0^\infty A(v)\{v^2-\frac{v}{Q_0(v)}\}dv, \tag{2.2.36}$$

so that

$$Q_1(u)=\frac{1}{A(u)}\int_u^\infty A(v)\{\frac{v}{Q_0(v)}-v^2\}dv. \tag{2.2.37}$$

The asymptotic behaviour of Q_1 for large u yields

$$\nu^{-1}Q_1(\nu^{1/3}s)=\frac{1}{4}\nu^{-1}+O(\nu^{-3}). \tag{2.2.38}$$

Likewise it is found that

$$\nu^{-5/3}Q_2(\nu^{1/3}s)=-\frac{1}{8}s\nu^{-4/3}+O(\nu^{-10/3}). \tag{2.2.39}$$

Asymptotic solution for region C. For $u\to-\infty$ the first two terms of (2.2.24) are

$$Q_0=u^2+\alpha+O(u^{-1}), \tag{2.2.40a}$$

$$Q_1 = \frac{1}{3u^3} - \frac{2}{3} \ln u + O(1), \tag{2.2.40b}$$

where $-\alpha$ denotes the largest zero of the Airy function From (2.2.40) it is seen that an asymptotic expansion for region C must contain fractional powers of ν as well as logarithmic terms. We assume that the solution has the formal power series expansion

$$p = f_0(x;\nu)\nu + \sum_{n=1}^{\infty} f_n(x;\nu)\nu^{(1-2n)/3} \tag{2.2.41}$$

with coefficients f_j that may still depend upon ln ν. From the recurrent system of equations for f_j it follows that

$$f_0 = a_0(\nu) + x - \frac{1}{3}x^3, \quad f_1 = a_1(\nu). \tag{2.2.42}$$

Because of (2.2.40) we have that $a_0(\nu) = 2/3$ and $a_1(\nu) = \alpha$. The next coefficient is

$$f_2 = \frac{-1}{x+1} + \frac{2}{3}\{\ln(2-x) - \ln(x+1)\} + a_2(\nu). \tag{2.2.43}$$

Matching the asymptotic solutions for regions B and C. We substitute

$$x = -1 + q\nu^{-1/3}$$

into (2.2.41) and

$$u = -q\nu^{1/3}$$

in (2.2.24). First the functions $Q_k(u)$ are analyzed for large negative values of u. The function

$$u = -Ai'(z)/Ai(z)$$

has a simple pole at $z = -\alpha$ with residue -1. Thus for z near $-\alpha$ we have

$$z \approx -\alpha - 1/u,$$

so that

$$\nu^{-1/3}Q_0(-a\nu^{1/3}q) = q^2\nu^{1/3} - q^{-1}\nu^{-2/3} + O(\nu^{-4/3}). \tag{2.2.44}$$

Using the relation

$$A(u)\{\frac{1}{3}u^3 - \frac{1}{3}\ln(u^2+c)\} = \tag{2.2.45}$$

$$-\int_u^{\infty} A(v)\{v^2 - \frac{v^3}{3Q_0^2} - \frac{2v}{3(v^2+c)} + \frac{\ln(v^2+c)}{3Q_0^2}\}dv,$$

one easily finds

$$\nu^{-1}Q_1(-\nu^{1/3}q) = -\frac{1}{3}q^3 + (b_1 - \frac{2}{9}\ln\nu + \tag{2.2.46}$$

$$-\frac{2}{3}\ln|q|)\nu^{-1}+O(\nu^{-5/3})$$

with

$$b_1=\frac{1}{A(-\infty)}\int_{-\infty}^{\infty}A(\nu)\{\frac{\nu}{Q_0}-\frac{\nu^3}{3Q_0^2}-\frac{2\nu}{3(\nu^2+1/2\alpha)}+ \tag{2.2.47}$$

$$+\frac{\ln(\nu^2+1/2\alpha)}{3Q_0^2}\}\,d\nu$$

On the other hand, the coefficients of the expansion (2.2.41) become

$$\nu f_0 = q^2\nu^{1/3}-\frac{1}{3}q^3,\quad \nu^{-1/3}f_1=\alpha\nu^{-1/3}, \tag{2.2.48}$$

$$\nu^{-1}f_2 = -q^{-1}\nu^{-2/3}+\{-\frac{2}{3}\ln|q| + \tag{2.2.49}$$

$$+\frac{2}{9}\ln\nu+\frac{2}{3}\ln 3+a_2(\nu)\}\nu^{-1}+O(\nu^{-4/3}).$$

From (2.2.43) and (2.2.49) we conclude that

$$a_2(\nu)=b_1-\frac{4}{9}\ln\nu-\frac{2}{3}\ln 3. \tag{2.2.50}$$

Asymptotic solution for region D. The expansion (2.2.41) breaks down as x approaches the value 2. Then $p=O(\nu^{-1})$ and because of (2.2.42)

$$x=2+\frac{1}{3}\alpha\nu^{-4/3}+O(\nu^{-2}).$$

We consider the inverse $x=x(p)$ and introduce the independent variable r:

$$p=r\nu^{-1}-\frac{2}{3}\nu^{-1}+\frac{5}{27}\alpha\nu^{-7/3}. \tag{2.2.51}$$

Eq.(2.2.17) changes into

$$-\{(r-\frac{2}{3}+\frac{5}{27}\alpha\nu^{-4/3})(x^2-1)+x\}\frac{dx}{dr}= \tag{2.2.52}$$

$$(r-\frac{2}{3}+\frac{5}{27}\alpha\nu^{-4/3})\nu^{-2}.$$

The specific transformation (2.2.51) has some computational advantages in a later stage. It is assumed that $x(r;\nu)$ can be expanded as

$$x=2+\frac{1}{3}\alpha\nu^{-4/3}+\sum_{j=2}^{\infty}\chi_j(r;\nu)\nu^{-2(j+1)/3}, \tag{2.2.53}$$

giving a recurrent system for χ_j with

$$\chi_2(r;\nu)=-\frac{1}{3}r+\frac{2}{9}\ln|r|+d_2(\nu). \tag{2.2.54}$$

Moreover, one computes that

$$\chi_3(r;\nu)=-\frac{2}{27}r^2+\cdots. \tag{2.2.55}$$

Matching of the asymptotic solutions for regions C and D. We choose as an intermediate coordinate χ:

$$x=2-\chi\nu^{-1}. \tag{2.2.56}$$

Substitution in (2.2.41) yields after reordering the expansion

$$p=3\chi+\alpha\nu^{-1/3}+(-2\chi^2-\frac{1}{3}+\frac{2}{3}\ln|\chi|-\frac{10}{9}\ln\nu$$
$$-\frac{4}{3}\ln 3+b_1)\nu^{-1}+O(\nu^{-4/3}). \tag{2.2.57}$$

On the other hand, substitution of

$$r=p\nu+\frac{2}{3}-\frac{5}{27}\alpha\nu^{-4/3} \tag{2.2.58}$$

in (2.2.53) gives

$$\chi=\frac{1}{3}p-\frac{1}{3}\alpha\nu^{-1/3}+\{\frac{2}{9}-\frac{2}{9}\ln|p|-\frac{2}{9}\ln\nu+$$
$$+\frac{2}{27}p^2-d_2(\nu)\}\nu^{-1}+O(\nu^{-4/3}). \tag{2.2.59}$$

These two expansions have to represent the same intermediate solution. Consequently, subsitution of one of them into the other produces an identity. This identity holds for

$$d_2(\nu)=\frac{1}{9}-\frac{16}{27}\ln\nu-\frac{2}{3}\ln 3+\frac{1}{3}b_1. \tag{2.2.60}$$

Asymptotic solution for region E. The asymptotic expansion (2.2.53) does not hold near $r=0$. In order to anlyse the local behaviour of the solution near this value, we introduce the independent variable

$$\xi=(x-x_s)\nu^2 \tag{2.2.61}$$

with x_s such that the exact periodic solution takes the value

$$p_s=-\frac{2}{3}\nu^{-1}+\frac{5}{27}\alpha\nu^{-7/3}. \tag{2.2.62}$$

Let

$$p=-\frac{2}{3}\nu^{-1}+\frac{5}{27}\alpha\nu^{-7/3}+\sum_{j=2}^{\infty}\eta_j(\xi,\nu)\nu^{-(5+2j)/3}, \tag{2.2.63}$$

$$x_s=2+\frac{1}{3}\alpha\nu^{-4/3}+\sum_{j=2}^{\infty}x_j(\nu)\nu^{-2(j+1)/3}. \tag{2.2.64}$$

Then the η_j follow from a recurrent system of equations by substituting (2.2.61)-(2.2.64) in (2.2.17) and equating equal powers of ν. The system starts with

$$\frac{d\eta_2}{d\xi}-\frac{9}{2}\eta_2+\frac{5}{2}(x_2+\xi)=0 \tag{2.2.65}$$

with $\eta_2(0;\nu)=0$, so

$$\eta_2(\xi;\nu)=\frac{5}{9}\xi+\{\frac{10}{81}+\frac{5}{9}x_2(\nu)\}\{1-\exp(\frac{9}{2}\xi)\}. \tag{2.2.66}$$

Matching the asympototic solutions for regions D and E. As the intermediate coordinate we introduce $s=r\nu$. Other choices would leave us with non-matchable exponential terms. In the intermediate region expansion (2.2.53) changes into

$$x(s;\nu)=2+\frac{1}{3}\alpha\nu^{-4/3}+\{\frac{2}{9}\ln\nu+d_2(\nu)\}\nu^{-2}+O(\nu^{-8/3}). \tag{2.2.67}$$

In order to bring the expansion (2.2.63) of region E in the required intermediate form we set

$$\xi=\sigma+2/9\ln\nu, \tag{2.2.68}$$

so that

$$s=\{-\frac{10}{81}-\frac{5}{9}x_2(\nu)\}\exp(\frac{9}{2}\sigma)+O(\nu^{-2/3}).$$

Inserting this expression in (2.2.67) we must get an identity as in the intermediate region the domains of asymptotic convergence of the two expansions overlap. This identity produces for $x_2(\nu)$ the transcendental equation

$$x_2(\nu)=-\frac{28}{29}\ln\nu+\frac{1}{9}-\frac{2}{3}\ln 3+ \tag{2.2.69}$$
$$+\frac{1}{3}b_1+\frac{2}{9}\ln|\frac{10}{81}+\frac{5}{9}x_2(\nu)|.$$

Matching the asymptotic solutions of regions E and A′. Substitution of

$$x=2+\frac{1}{3}\alpha\nu^{-4/3}+\{x_2(\nu)+\xi\}\nu^{-2}+\cdots \tag{2.2.70}$$

into the expansion (2.2.21) for region A' yields

$$p=-\frac{2}{3}\nu^{-1}+\frac{5}{27}\alpha\nu^{-4/3}+ \tag{2.2.71}$$
$$+[\frac{5}{9}\{x_2(\nu)+\xi\}+\frac{10}{81}]\nu^{-3}+O(\nu^{-11/3}),$$

which agrees with (2.2.63) for $\xi\to-\infty$. Thus, we now completed our construction of locally valid matched asymptotic approximations of the periodic solution.

Amplitude and period. Substitution of

$$r=\frac{2}{3}-\frac{5}{27}\alpha\nu^{-4/3} \tag{2.2.72}$$

in (2.2.62) yields the amplitude a_ν of the periodic solution:

$$\begin{aligned} a_\nu = {} & 2+\frac{1}{3}\alpha\nu^{-4/3} + \\ & +(\frac{1}{3}b_1-\frac{16}{27}\ln\nu-\frac{1}{9}+\frac{2}{9}\ln 2-\frac{8}{9}\ln 3)\nu^{-2}+ \\ & +(\frac{1}{3}b_2-\frac{2}{27}\alpha^2)\nu^{-8/3}+ \\ & +(\frac{1}{3}b_3+\frac{104}{243}\alpha\ln\nu-\frac{4}{27}\alpha b_1-\frac{91}{486}\alpha+ \\ & +\frac{52}{81}\alpha\ln 3-\frac{13}{81}\alpha\ln 2)\nu^{-10/3}+O(\nu^{-4}\ln^2\nu) \end{aligned} \tag{2.2.73}$$

with

$$\alpha=2.33810741, \quad b_1=.17235, \tag{2.2.74}$$

$$b_2=.61778, \quad b_3=-.55045. \tag{2.2.75}$$

The last two terms have been computed in Bavinck and Grasman (1974).

The period is found from

$$T=2\int_{-a_\nu}^{a_\nu}\frac{dx}{p(x)}. \tag{2.2.76}$$

For each region a different local approximation for p holds. The integration in a region is taken over an x-interval that extends in both directions to a point in the domain of overlap with the adjacent regions. In region D the integration is over p, so the contribution to the period is

$$T_D=\int_D\frac{1}{p}\frac{dx}{dp}dp=-\int_D\frac{dp}{\nu(x^2-1)p+x}. \tag{2.2.77}$$

The end points are arbitrarily chosen in the domains of overlap. In the final summing their values cancel. The final result reads

$$\begin{aligned} T = {} & (3-2\ln 2)\nu+3\alpha\nu^{-1/3}-\frac{2}{3}\nu^{-1}\ln\nu+\{\ln 2-\ln 3+ \\ & +3b_1-1-\ln\pi-2\ln Ai'(-\alpha)\}\nu^{-1}+O(\nu^{-4/3}\ln\nu). \end{aligned} \tag{2.2.78}$$

2.2.3 The Lienard plane

By the method of Waldvogel (1985) only three local asymptotic solutions are needed to approximate the periodic solution of (2.2.1). The equation is rewritten as

$$\epsilon\frac{dx}{dt} = -y+x-\frac{1}{3}x^3, \tag{2.2.79a}$$

$$\frac{dy}{dt} = x. \tag{2.2.79b}$$

Since we have to construct an approximation in the Lienard plane and to compute the period, we consider the equations

$$(x-\frac{1}{3}x^3-y)\frac{dy}{dx} = \epsilon x, \tag{2.2.80a}$$

$$\frac{dt}{dx} = \frac{1}{x}\frac{dy}{dx}. \tag{2.2.80b}$$

Near the stable branch of $y=x-\frac{1}{3}x^3$ with $x<0$ the solution is expanded as

$$y(x) = y_0(x)+\epsilon y_1(x)+\epsilon^2 y_2(x)+\cdots \tag{2.2.81}$$

$$y_0(x) = x-\frac{1}{3}x^3,\ldots. \tag{2.2.82}$$

Substitution in (2.2.82b) yields

$$t = t_0(x)+\epsilon t_1(x) + \epsilon^2 t_2(x)+\cdots, \tag{2.2.83}$$

$$t_0(x) = \ln(-x)-\frac{1}{2}x^2, \tag{2.2.84a}$$

$$t_1(x) = \ln\sqrt{1-1/x^2}+1/(x^2-1),\ldots. \tag{2.2.84b}$$

Rearranging the expansion (2.2.83), we obtain

$$t(x) = \ln(-x)-\frac{1}{2}x^2+(\epsilon+\epsilon^2+2\epsilon^3+5\epsilon^4+\cdots)\ln\sqrt{1-1/x^2}+$$

$$+(\epsilon+\frac{1}{2}\epsilon^2+\epsilon^3+\frac{5}{2}\epsilon^4+\cdots)/(x^2-1)+\cdots. \tag{2.2.85}$$

The complete series in ϵ is found from substitution of

$$y(x) = xf(\epsilon) + O(x^3) \tag{2.2.86}$$

in (2.2.80a):

$$f(\epsilon)(1-f(\epsilon)) = \epsilon. \tag{2.2.87}$$

When x approaches the value -1, the local variables ξ and η are introduced:

$$x = -1+\epsilon^{1/3}\xi, \quad y = -\frac{2}{3}+\epsilon^{2/3}\eta, \tag{2.2.88}$$

so that (2.2.80a) transforms into

$$\epsilon\frac{d\eta}{d\xi}(\xi^2-\eta-\frac{1}{3}\epsilon^3) = \epsilon(-1+\epsilon^{1/3}\xi) \tag{2.2.89}$$

with an asymptotic solution of the form

$$\eta(\xi) = \eta_0(\xi) + \epsilon^{1/3}\eta_1(\xi) + \cdots \tag{2.2.90}$$

Eq. (2.2.80) transforms into a similar equation with a solution that is expanded as

$$t(\xi;\epsilon) = C_1(\epsilon) + \epsilon^{2/3}T_0(\xi) + \epsilon T_1(\xi) + \cdots \tag{2.2.91}$$

For the coefficients $\eta_j(\xi)$ and $T_j(\xi)$ a recurrent system of equations holds which can be solved while matching with (2.2.81) and (2.2.83) is taken into account. The procedure is the same as the one of the preceding section. The equation for η_0 is solved implicity with $\xi = \xi_0(\eta)$ in terms of Airy functions. For $C_1(\epsilon)$ we obtain a power series in ϵ with coefficients depending on $\ln\epsilon$. Once the solution has left a neighbourhood of $(x,y) = (-1,-2/3)$, it enters a state of rapid change in which y is almost constant and x tends to the value 2. We make the following transformations:

$$\begin{aligned} x &= -1+u, \\ y &= -\frac{2}{3} + \alpha\epsilon^{2/3} + \epsilon v, \\ t &= \epsilon\tau \end{aligned} \tag{2.2.92}$$

with $-\alpha$ the first zero of the Airy function and use the equations for $du/d\tau$ and $dv/d\tau$ to compute the coefficients in

$$v = v_0(u_0) + \epsilon^{2/3}v_1(u_0) + \epsilon v_2(u_0) + \cdots, \tag{2.2.93a}$$

$$\tau = C_2(\epsilon)/\epsilon + \tau_0(u_0). \tag{2.2.93b}$$

Matching with (2.2.91) yields an asymptotic series in ϵ for $C_2 - C_1$. As u tends to 3, the asymptotic solution approaches the stable branch of $y = x - 1/3x^3$ with $x > 0$. The approximation of the solution is similar to (2.2.81). For t as a function of x we obtain

$$t = C_3(\epsilon) + \ln x - \frac{1}{2}x^2 + \epsilon\{\ln\sqrt{1-x^2} + 1/(x^2-1)\} + \epsilon^2(\ln\sqrt{1-x^2} + \cdots) + \cdots \tag{2.2.94}$$

with $C_3 - C_2$ a series in ϵ with coefficients determined by matching.

The time needed to go from a point x^* at the negative stable branch to its reflection with respect to the origin equals half the period of the solution:

$$\frac{1}{2}T = t(-x^*) - t(x^*). \tag{2.2.95}$$

It is easily verified that

$$\frac{1}{2}T = C_3 = (C_3 - C_2) + (C_2 - C_1) + C_1. \tag{2.2.96}$$

Because of the property (2.2.87) the coefficients of the series for $C_3 - C_2, C_2 - C_1$ and C_1 can be computed upto arbitrary order. Waldvogel (1985) used a formula manipulation package (SYMBAL) to carry out this computation.

2.2.4 Approximations of amplitude and period

The periodic solution of the Van der Pol equation

$$\frac{d^2x}{dt^2} + \mu(x^2-1)\frac{dx}{dt} + x = 0, \ \mu>0 \tag{2.2.97}$$

has been approximated in various ways. For $0<\mu<<1$ a regular perturbation technique can be applied, see Minorsky (1962). This method results in power series expansions for amplitude and period:

$$a = 2 + \frac{1}{96}\mu^2 - \frac{1033}{552960}\mu^4 + \cdots, \tag{2.2.98}$$

$$T = 2\pi(1 + \frac{1}{16}\mu^2 - \frac{5}{3072}\mu^4 + \cdots), \tag{2.2.99}$$

ν	T_{num}	T_{as}	a_{num}	a_{as}
1	6.66329	7.30478	2.00862	2.41212
2	7.62987	7.90201	2.01989	2.09527
3	8.95909	9.01936	2.02330	2.04622
4	10.20352	10.31171	2.02296	2.03232
5	11.61223	11.69129	2.02151	2.02608
10	19.07837	19.10699	2.01428	2.01474
15	26.82575	26.84118	2.01020	2.01031
20	34.68232	34.69219	2.00779	2.00738
25	42.59579	42.60274	2.00624	2.00626
30	50.54369	50.54890	2.00516	2.00517
40	66.50137	66.50466	2.00379	2.00379
50	82.50833	82.51064	2.00296	2.00296
60	98.54479	98.54651	2.00240	2.00240
70	114.60067	114.60203	2.00201	2.00201
80	130.67020	130.67130	2.00172	2.00172
90	146.49795	146.75068	2.00150	2.00150
100	162.83707	162.83782	2.00132	2.00132

Numerical and asymptotic approximation of period T and amplitude a of the Van der Pol oscillator for different values of ν

Table 2.2.1

see Clenshaw (1966), Deprit and Rom (1967), Anderson and Geer (1982) and Dadfar *et al.*(1984). The use of the symbolic manipulation system MACSYMA in the last two papers made it possible to compute the coefficients upto the terms of order $O(\mu^{24})$. Employing the regularity in the sequence of coefficients, the authors also computed the coeficients upto $O(\mu^{164})$. Moreover, a transformation of the parameter was made such that the singularity in the complex μ-plane moved to infinity. In this way the radius of convergence of the series (2.2.98) tends to infinity.

Numerical approximations of the periodic solution and its amplitude and period have been made with the following numerical methods: preditor-corrector (Urabe, 1957, 1958, 1960ab and 1963), modified Runge-Kutta (Krogdahl, 1960), Nordsieck (Ponzo and Wax, 1965c), approximation by Lie series (Maess, 1965), approximation by Tchebychev series (Clenshaw, 1966) and Runge-Kutta with variable stepsize (Zonneveld, 1966), see also Yanagiwara (1960). In table 2.2.1 we give numerical values for amplitude and period for $1<\mu<100$. These are based on the results of Zonneveld (1966) and Strasberg (1973). In the table they are compared with the asymptotic value obtained from (2.2.73) and (2.2.78).

For the quasi-periodically forced Van der Pol equation

$$\frac{d^2x}{dt^2}+\lambda(x^2-1)\frac{dx}{dt}+x=a\cos\nu_1 t+b\cos\nu_2 t \qquad (2.2.100)$$

solutions are approximated by Galerkin methods in Mitsui (1977) and in Shinohara *et al.*(1984).

Exercises

2.2.1 Use instead of (2.2.3) the stretching transformation

$$t=\xi\epsilon^{\alpha}, \quad x=1+\epsilon^{\beta}v(\xi)$$

and find out why $\alpha=2/3$ and $\beta=1/3$.

2.2.2 Show that $v_0=z'(\xi)/z(\xi)$ given by (2.2.6) satisfies $v_0\approx\sqrt{-\xi}$ for $\xi\to-\infty$, only if $B=C=0$.

2.2.3 Compute the first terms in the asymptotic expression for the period of the Van der Pol oscillator from the solution in the Lienard plane. Try to improve (2.2.78).

2.2.4 Show that the asymptotic formula (2.2.78) for the period of the Van der Pol oscillator is identical to the one found by Urabe (1960a):

$$T=(3-2\ln 2)\nu+3\alpha\nu^{-1/3}-\frac{2}{3}\nu^{-1}\ln\nu$$

$$+(3\ln 2-\ln 3-\frac{3}{2}+b_1-2d)\nu^{-1}+O(\nu^{-4/3}),$$

where

$$d=\frac{1}{2}\ln\alpha-b_1-\frac{1}{4}+\int_{-\infty}^{0}\{\frac{u}{Q_0(u)}-\frac{u}{u^2+\alpha}\}du+$$
$$+\int_{0}^{\infty}\{Q_0(u)-\frac{u^2}{2(u^3+1)}\}du,$$

see Veling (1983).

2.3 THE VOLTERRA-LOTKA EQUATIONS

In this section we deal with a type of oscillation, which differs substantially from the Van der Pol relaxation oscillator. The Volterra-Lotka equations have a one parameter family of periodic solutions enclosing a center point of the system. In section 2.3.1 the biological meaning of these equations is reviewed. In sections 2.3.2 - 2.3.4 we give asymptotic approximations of solutions with a large amplitude. In particular an asymptotic expression for the period is derived.

2.3.1 Modeling prey-predator systems

For the quantitative analysis of populations use is made of mathematical models such as differential equations describing the dynamics of interacting populations. Let us consider the following predator-prey model

$$\frac{dx}{dt} = ax-bxy-ex^2, \tag{2.3.1a}$$

$$\frac{dy}{dt} = -cy+dxy, \tag{2.3.1b}$$

where x and y denote the densities of a prey and a predator species living in a same environment. All constants are positive. In the absence of predation ($y=0$) the prey density (2.3.1a) would satisfy the well-known logistic differential equation and x would tend to the stationary state $x=a/e$ as $t\rightarrow\infty$. When there is predation the system tends to the equilibrium

$$(x,y) = (\frac{c}{d}, \frac{ad-ec}{bd}), \quad \text{if } ad>ec,$$

$$(x,y) = (\frac{a}{e},0), \quad \text{if } ad<ec.$$

There are many examples in ecology of populations with densities fluctuating with a more or less fixed period. The example of the Canadian lynx and snowshoe hare is classical, see fig.2.3.1. It was the fluctuation of densities of two fish species having a predator-prey relation that led Volterra (1931) to a model of type (2.2.1) with $e=0$, see also Lotka (1925). Note that for $e>0$ such fluctuations tend to damp out.

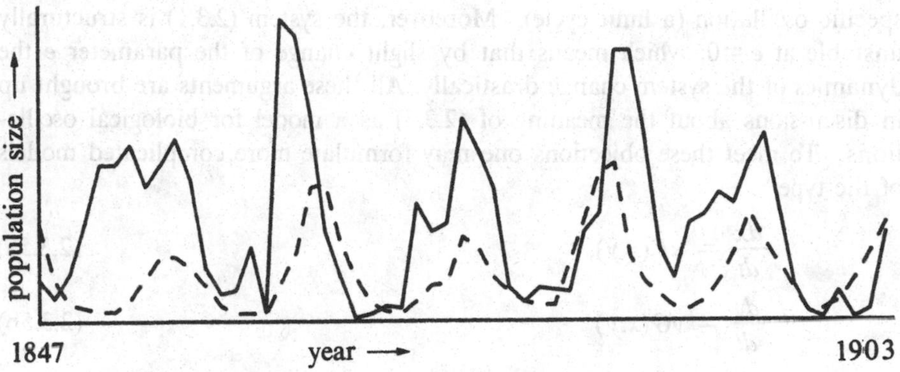

Fig.2.3.1 Fluctuations in the populations of the Canadian lynx (solid line) and the snowshoe hare (dotted line) from the trading figures of the Hudson Bay Company. It is noted that peaks in the hare population are followed by peaks in the lynx population.

For convenience we transform the system (2.3.1) with $e=0$ by using the substitutions

$$x^{*}=\frac{d}{c}x, \quad y^{*}=\frac{b}{a}y, \quad \nu=\frac{a}{c}, \quad t^{*}=-at. \tag{2.3.2}$$

Then we will find the following system, where we have dropped the asterisks and which we will use as standard form from now on:

$$\frac{dx}{dt} = x(1-y), \quad x(0)=\theta, \quad 0<\theta<1, \tag{2.3.3a}$$

$$\nu\frac{dy}{dt} = y(-1+x), \quad y(0)=1. \tag{2.3.3b}$$

In the system (2.3.3) we have also specified the initial conditions. Eq.(2.3.3) can be transformed into a Hamiltonian system, see Dutt (1976). Then

$$H = x - \ln x + \nu(y - \ln y) - 1 - \nu \qquad (2.3.4)$$

is the conserved quantity ($dH/dt = 0$) and $H = \theta - \ln \theta - 1$. The periodic solutions form a one parameter family with the equilibrium $(x,y) = (1,1)$ as center point. The period of such a solution depends heavily on the initial data which for biological systems is quite unnatural. Any disturbance may bring the system in a different orbit: there is no mechanism that drives it back to one specific oscillation (a limit cycle). Moreover, the system (2.3.1) is structurally unstable at $e = 0$; which means that by slight change of the parameter e the dynamics of the system change drastically. All these arguments are brought up in discussions about the meaning of (2.3.3) as a model for biological oscillations. To meet these objections one may formulate more complicated models of the type

$$\frac{dx}{dt} = xF(x,y), \qquad (2.3.5a)$$

$$\frac{dy}{dt} = yG(x,y) \qquad (2.3.5b)$$

and carry out a qualitative analysis of such system in order to detect the existence of limit cycles. Nevertheless we have the idea that a study of the model equations (2.3.3) might be very useful in spite of the above objections. If for a prey-predator system we can find parameters $a,...,d$ such that H is almost constant, then the equations (2.3.3) form a good first approximation of the biological problem. With perturbation methods one may construct higher approximations of the full problem. This technique has proved to be succesful in the theory of almost linear oscillations, where the harmonic oscillator (also a conservative system) forms the first order approximation.

For H small the initial value θ is close to 1. This choice corresponds with a small amplitude oscillation around the equilibrium, which is covered by the theory of almost linear oscillations, see e.g. Bogoliubov and Mitropolsky (1961). In a first order approximation this oscillation is sinusoidal with period $2\pi\sqrt{\nu}$.

In the next section we consider large amplitude oscillations, as analyzed by Lauwerier (1975). In section 2.3.3 we consider periodic solutions of a singular perturbed Volterra-Lotka system. Finally, in section 2.3.4 a method is given to compute the period by inverse Laplace asymptotics. The period appears to increase monotonicly with the distance of the closed curve to the equilibrium point, see Rothe (1984) and Waldvogel (1984).

2.3.2 Oscillations with both state variables having a large amplitude

A different type of asymptotic approximation applies to the case where $H >> 1$ (large perturbation of the equilibrium). Substitution of

$$x = H\xi, \quad y = H\eta \tag{2.3.6}$$

will give a closed trajectory in the ξ,η-plane around $(1/H, 1/H)$, that tends to a triangle as $H \to \infty$ see fig. 2.3.2. The sides of the triangle are $\xi=0$, $\eta=0$, and $\xi+\nu\eta=1$. Since

$$dt = \int \frac{d\xi}{\xi(1-H\eta)} = \int \frac{-d\eta}{\eta(1-H\xi)}, \tag{2.3.7}$$

we expect the main contribution to the period from the parts of the trajectory along the ξ- and η-axis, where we have the reduced equations

$$\frac{d\xi_0}{dt} = \xi_0 \text{ and } \frac{d\eta_0}{dt} = -\frac{\eta_0}{\nu}. \tag{2.3.8}$$

According to (2.3.4) ξ takes a minimal value for $t = t_B$ with

$$\xi_{\min} = H^{-1}e^{-H+\nu} + O(H^{-1}e^{-2H+2\nu}). \tag{2.3.9}$$

at $\eta = 1/H$. Along the ξ-axis we then have

$$\xi_0(t) = H^{-1}e^{-H+\nu}e^{t-t_B}. \tag{2.3.10}$$

This approximation is valid as long as $\eta = O(1/H)$, which is the case for $\xi_{\min} < \xi < \xi_C$, with $1/H < \xi_C < 1$. From (2.3.10) we conclude that

$$t_C - t_B = H + \ln H + O(1). \tag{2.3.11}$$

Along the remaining part of the η-axis we have

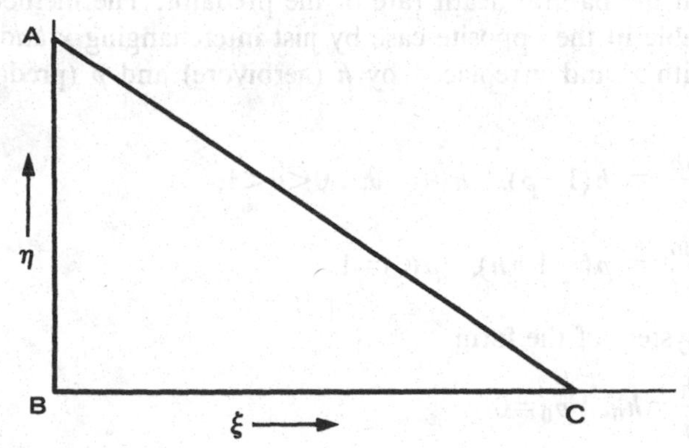

Fig.2.3.2 The limit case $H \to \infty$. The closed trajectory given by (2.3.4) tends to a triangle.

$$\eta_0(t)=H^{-1}e^{-(t_C-t)/H\nu} \tag{2.3.12}$$

Likewise this approximation is valid as long as $\xi=O(1/H)$, which is the case for $1/H<\eta<\eta_A$, with $\eta_A<1/\nu$. From (2.3.12) we conclude that

$$t_B-t_A=\nu \ln H+O(1). \tag{2.3.13}$$

So we have found

$$T\sim t_C-t_A=H+(1+\nu)\ln H+O(1). \tag{2.3.14}$$

Lauwerier (1975) constructed higher order approximations for this problem, showing that (2.3.14) is indeed the main contribution to the period. He derived

$$t=H+(1+\nu)\ln H-\nu\ln\nu+O(H^{-1}\ln H). \tag{2.3.15}$$

Thus the period is in a first order approximation proportional with the parameter H of the system, which is supposed to be large.

It is noted that strictly speaking the oscillation cannot be qualified as a relaxation oscillation as defined in section 1.1.2. The relaxation parameter H follows from the initial values, which makes it slightly different from the usual type of relaxation oscillation, where the relaxation parameter is in the coefficients of the system of differential equations.

2.3.3 Oscillations with one state variable having a large amplitude

In the sequel we assume that ν is a small parameter and replace ν by ϵ (Grasman and Veling, 1973). It means that the natural growth rate of the prey is smaller than the natural death rate of the predator. The method we discuss is also applicable in the opposite case by just interchanging x and y. The system (2.3.3) with x and y replaced by h (herbivore) and p (predator) has the form

$$\epsilon\frac{dh}{dt}=h(1-p),\quad h(0)=\theta,\quad 0<\theta<1, \tag{2.3.16a}$$

$$\epsilon\frac{dp}{dt}=p(-1+h),\quad p(0)=1. \tag{2.3.16b}$$

The reduced system of the form

$$\frac{dh}{dt}=h_0,\quad p_0=0 \tag{2.3.17}$$

does not satisfy the initial value for p. Therefore, we expect a boundary layer behaviour in an ϵ-dependent time interval where p decreases from the value 1 to 0. For the prey density we have for $p=0$:

$$h_0(t)=\theta e^t. \tag{2.3.18}$$

However, there is a limitation in the exponential growth of the prey. According to

$$h - \ln h + \epsilon(p - \ln p) = \theta - \ln \theta + \epsilon \tag{2.3.19}$$

there is a maximum at $h_0(T_0) = \mu$, with μ and T_0 satisfying

$$\mu - \ln \mu = \theta - \ln \theta, \quad T_0 = \mu - \theta. \tag{2.3.20}$$

When h is at its maximum, p must be 1. Apparently there is also a boundary layer in an ϵ-dependent time interval at $t = T_0$, where p increases from a value near 0 to 1. From (2.3.16b) we see that p continues to increase rapidly until h arrives in an ϵ-neighbourhood of the line $h=1$, where $p = O(1/\epsilon)$. From (2.3.16a) it is seen that at this stage h decreases swiftly, so that in turn p starts to decrease rapidly as soon as h has passed the ϵ-neighbourhood of $h=1$. This brings us back at the initial point $(h,p) = (\theta, 1)$, see fig.2.3.3.

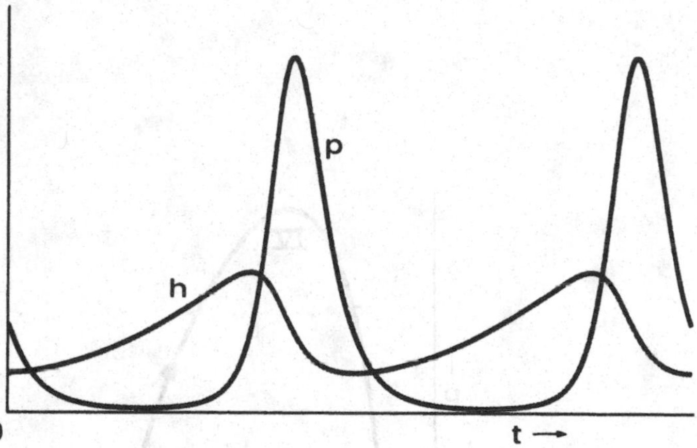

Fig.2.3.3 The fluctuations in the densities of herbivore (h) and predator (p) population for the system (2.3.16) with $\epsilon = .1$ and $\theta = .5$. When the density of the predator is low, the growth of the number of herbivores is almost exponential. The outbursts of predators bring the herbivores at the starting level.

By heuristic reasoning we have derived the following limit behaviour of the system as $\epsilon \to 0$: repeatedly the prey density increases exponentially from θ to μ in a time T_0, then there is an outburst of predators eating away the prey species upto the starting level θ. The predator density as a function of time has the shape of periodic pulses at times nT_0, $n = 1,2,\ldots$.

We will write the solution in the phase plane explicitly as $p = p(h;\epsilon)$ or $h = h(p;\epsilon)$. Therefore, we have to distinguish four separate regions as sketched in fig.2.3.4. Within these regions the solution of (2.3.19) is unique. The implicit function theorem is used for the qualitative and quantitative analysis of the problem. For each region the solution has a convergent power series expansion. This can be proved from theorem 2.3.1 for ϵ sufficiently small. The regions are separated by the points

$$h_1 = \theta(1+\frac{1}{\theta-1}\epsilon \ln \epsilon + \epsilon \xi_1), \quad p_1 = \epsilon\eta_1, \tag{2.3.21a}$$

$$h_2 = \mu(1+\frac{1}{\mu-1}\epsilon \ln \epsilon - \epsilon \xi_2), \quad p_2 = \epsilon\eta_2, \tag{2.3.21b}$$

$$h_3 = \mu(1-\xi_3), \quad p_3 = \eta_3/\epsilon, \tag{2.3.21c}$$

$$h_4 = \theta(1+\xi_4), \quad p_4 = \eta_4/\epsilon. \tag{2.3.21d}$$

The two theorems we need are formulated as follows.

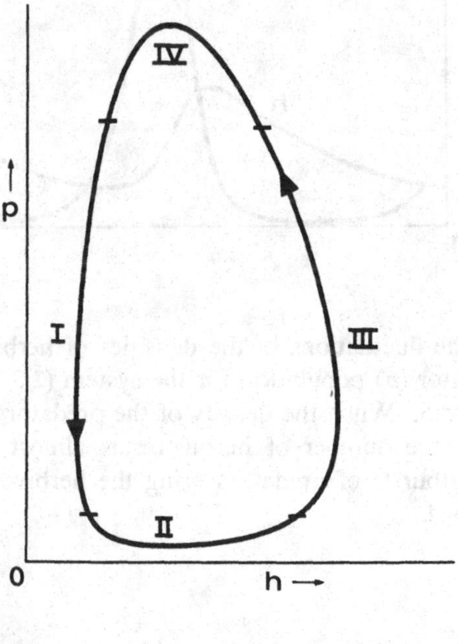

Fig.2.3.4 The four segments of the closed curve representing the periodic solution in the phase plane. The seperation points satisfy (2.3.21) and (2.3.41). In the computation of the period their values cancel.

Theorem 2.3.1 Let $w=f(z)$ be regular in a neighbourhood of $z=z_0$ and $w_0=f(z_0)$. A necessary and sufficient condition for $w=f(z)$ to have a unique solution $z=F(w)$, being regular in a neighbourhood of $w=w_0$, is $f'(z_0)\neq 0$. This solution exists if

$$|w-w_0|<\frac{R^2a^2}{6M} \text{ for } |z-z_0|\leqslant\frac{R^2a}{4M}, \tag{2.3.22}$$

where $R<R'$ is the convergence radius of the power series $f(z)$ around $z=z_0$, $a=|f'(z_0)|$ and

$$M=\max\{|f(z)| \text{ for } |z-z_0|\leqslant R\}. \tag{2.3.23}$$

The solution $z=F(w)$ can be written as

$$z=F(w)=\frac{1}{2\pi i}\int_C \frac{z\,f'(z)}{f(z)-w}dz, \tag{2.3.24}$$

where C is the circle at z_0 with radius $\frac{1}{4}Ra^2/M$.

Proof. See Copson (1948), Ch.6.22.

Theorem 2.3.2 Let $w=f(z)$ be regular in a neighbourhood of $z=z_0$ and let $f'(z_0)\neq 0$, then there exists a unique function

$$F(w)=z_0+\sum_{n=1}^{\infty} k_n(w-w_0)^n \tag{2.3.25}$$

regular in a neighbourhood of $w=w_0$. Moreover, $z=F(w)$ is the solution of $w=f(z)$ and

$$k_n=\frac{1}{n!}\left[\frac{d^{n-1}}{dz^{n-1}}\{\phi(z)\}^n\right]_{z=z_0}, \tag{2.3.26}$$

with $f(z)-w_0=(z-z_0)/\phi(z)$, $\phi(z)\neq 0$ in a neighbourhood of $z=z_0$.

Proof. See Copson (1948), Ch.6.23.

With the aid of these two theorems we determine the functions $p(h;\epsilon)$ and $h(p;\epsilon)$ for the four regions. We introduce the notation

$$L(z)=z-\ln z. \tag{2.3.27}$$

The results are as follows:

$$I:\quad h=\theta+\sum_{k=1}^{\infty} a_n\rho_1^n,\quad \rho_1=\epsilon(L(p)-1) \tag{2.3.28}$$

$$a_n=\frac{1}{n!}\left[\frac{d^{n-1}}{dx^{n-1}}\left\{\frac{x}{\ln(1+x/\theta)-x}\right\}^n\right]_{x=0},$$

$$a_1=\frac{\theta}{(1-\theta)},\quad a_2=\frac{\theta}{2(1-\theta)^3},$$

$$II: \quad p=\sum_{n=1}^{\infty}\frac{n^{n-1}}{n!}\rho_2^n, \quad \rho_2=\exp\{\frac{L(h)-L(\theta)-\epsilon}{\epsilon}\}, \tag{2.3.29}$$

$$III: \quad h=\mu+\sum c_n\rho_3^n, \quad \rho_3=\epsilon(L(p)-1), \tag{2.3.30}$$

$$c_n=\frac{!}{n!}[\frac{d^{n-1}}{dx^{n-1}}\{\frac{x}{\ln(1+x/\mu)-x}\}^n]_{x=0},$$

$$c_1=\frac{\mu}{(1-\mu)}, \quad c_2=\frac{\mu}{2(1-\mu)^3},$$

$$IV: \quad p=\rho_4+\ln\rho_4+ \tag{2.3.31}$$

$$\frac{\ln\rho_4}{\rho_4}\sum_{n=0}^{\infty}\sum_{m=0}^{n}d_{n,m}(\ln\rho_4)^{n-m}\rho_4^{-n},$$

$$\rho_4=\frac{L(\theta)-L(h)+\epsilon}{\epsilon}, \quad d_{0,0}=1, \quad d_{1,0}=\frac{1}{2}, \quad d_{1,1}=1,$$

$$d_{n,m}=\frac{1}{(n-m)!}\{\begin{bmatrix}n+1\\m\end{bmatrix}[\frac{d^{n-m}}{dx^{n-m}}\frac{e^x x^{n+2}}{(e^x-1)^{n+2}}]_{x=0}+$$

$$-\begin{bmatrix}n\\m-1\end{bmatrix}[\frac{d^{n-m}}{dx^{n-m}}\frac{x^{n+1}}{(e^x-1)^{n+1}}]_{x=0}\}.$$

For region II we will prove the correctness of the representation (2.3.29), for the other regions we refer to Veling (1973). From (2.3.19) we find

$$pe^{-p}=\exp\{\frac{L(h)-L(\theta)-\epsilon}{\epsilon}\}=\rho_2(h), \tag{2.3.32}$$

and for h in region II, it means $h_1<h<h_2$. We will calculate the $\max\{\rho_2(h)|h_1<h<h_2\}$. In view of the fact that ρ_2 is monotonicly decreasing for $h<1$ and monotonicly increasing for $h>1$, it suffices to calculate $\rho_2(h_1)$ and $\rho_2(h_2)$. We have

$$\rho_2(h_1)=\exp[\{L(\theta)+\frac{\theta}{\theta-1}\epsilon\ln\epsilon+\theta\epsilon\xi_1+ \tag{2.3.33}$$

$$-\ln(1+\frac{1}{\theta-1}\epsilon\ln\epsilon+\epsilon\xi_1)-L(\theta)-\epsilon\}/\epsilon].$$

or

$$\rho_2(h_1)=\epsilon\exp\{(\theta-1)\xi_1-1\}+O(\epsilon\ln^2\epsilon)=O(\epsilon), \tag{2.3.34}$$

and, likewise

$$\rho_2(h_2)=O(\epsilon). \tag{2.3.35}$$

Thus $\max\{\rho_2(h)\}=O(\epsilon)$, and also $\max\{p(h)\}=O(\epsilon)$. Consider now

$$w = f(z) = ze^{-z}, \; \phi(z) = z/w = e^z$$

with

$$z_0 = 0 \text{ and } f(z_0) = w_0 = 0.$$

Then $f(z)$ is regular for

$$z = z_0 = 0 \text{ and } f'(z_0) = f'(0) = 1 \neq 0.$$

According to Theorems 2.3.1 and 2.3.2 we find that $w = f(z)$ possesses a unique solution of the form

$$z = \sum_{n=1}^{\infty} b_n w^n, \quad |z| \leq \frac{R^2 a}{4M}, \quad |w| < \frac{R^2 a^2}{6M}, \tag{2.3.36}$$

with

$$R = 1 < R' \tag{2.3.37}$$

(convergence radius power series $f(z)$ is infinite),

$$M = \max\{|f(z)| \text{ for } |z| \leq R = 1\} = |f(-1)| = e^{-1}. \tag{2.3.38}$$

$$a = f'(z_0) = f'(0) = 1 \tag{2.3.39}$$

and

$$b_n = \frac{1}{n!}\left[\frac{d^{n-1}}{dz^{n-1}}(e^z)^n\right]_{z=0} = \frac{n^{n-1}}{n!}. \tag{2.3.40}$$

Confining z and w to real values (p and ρ_2), we find that the conditions of the theorem are satisfied if $p \leq e/4$ and $\rho_2 < e/6$, which can easily be achieved by taking ϵ sufficiently small. So we derived (2.3.39). Regions I, III are handled in a similar way. For region IV, however, we have to estimate first the part of the expansion that tends to infinity for $\epsilon \to 0$. The remaining part of the expansion can be treated in the same way as the other regions, where we need the representation of Theorem 2.3.1, see Veling (1973).

We now give the interdependence of ξ_i and η_i of (2.3.21)- (2.3.24). Since we know that (2.3.28)-(2.3.31) represent convergent expansions, we calculate the interdependence of ξ_1 and η_1 by substituting $h = h_1$ in ρ_1 giving p_1 to take one example. We find

$$\eta_1 = \exp\{(\theta - 1)\xi_1 - 1\} + O(\epsilon \ln^2 \epsilon), \tag{2.3.41a}$$

$$\eta_1 = \exp\{(1 - \mu)\xi_2 - 1\} + O(\epsilon \ln^2 \epsilon), \tag{2.3.41b}$$

$$\eta_3 = \ln(1 - \xi_3) + \mu \xi_3 + O(\epsilon \ln \epsilon), \tag{2.3.41c}$$

$$\eta_4 = \ln(1 + \xi_4) - \theta \xi_4 + O(\epsilon \ln \epsilon). \tag{2.3.41d}$$

From the explicit solutions of the four regions, information can be obtained about the periodic solution of the system. The period is determined by an integration along the periodic orbit consisting of four parts

$$T = T_I + T_{II} + T_{III} + T_{IV}, \tag{2.3.42}$$

with

$$T_I = \int_{p_4}^{p_1} \frac{dp}{(dp/dt)}, \quad T_{II} = \int_{h_1}^{h_2} \frac{dh}{(dh/dt)}, \tag{2.3.43}$$

$$T_{III} = \int_{p_2}^{p_3} \frac{dp}{(dp/dt)}, \quad T_{IV} = \int_{h_3}^{h_4} \frac{dh}{(dh/dt)}. \tag{2.3.44}$$

By the choice of the points $(h_i,p_i), i=1,2,3,4$, we were able to give expansions wich are valid in such a large domain that two solutions of adjacent regions are overlapping in a neighbourhood of the boundary points. So we can get all information about the solutions from the expansions. The main contribution to the period will come from T_{II}. We know $p=p(h)=O(\epsilon)$, so from (2.3.16a) and (2.3.43)

$$T_{II} = \ln\mu - \ln\theta + O(\epsilon\ln\epsilon) = \mu - \theta + O(\epsilon\ln\epsilon). \tag{2.3.45}$$

In a similar way it can be shown that

$$T_I = O(\epsilon\ln\epsilon), \quad T_{III} = O(\epsilon\ln\epsilon), \quad T_{IV} = O(\epsilon), \tag{2.3.46}$$

so

$$T = \mu - \theta + O(\epsilon\ln\epsilon). \tag{2.3.47}$$

Veling (1973) computed higher order terms for the asymptotic expression of the period. It should be noted that the terms with ξ_i and η_i arising in $T_I, T_{II}, T_{III}, T_{IV}$ cancel in the final summing of these four contributions to the period. The result reads

$$T = (\mu-\theta) + \epsilon\ln\epsilon\left(\frac{-1}{1-\theta} + \frac{1}{1-\mu}\right) + \tag{2.3.48}$$

$$+\epsilon\left[\frac{1}{1-\theta} - \frac{1}{1-\mu} + \frac{1}{1-\theta}\ln\{(1-\theta)\ln(\frac{1}{\theta})\} + \right.$$

$$\left. -\frac{1}{1-\mu}\ln\{(\mu-1)\ln\mu\} + I(\theta) + I(\mu)\right] + O(\epsilon^2\ln^2\epsilon)$$

with

$$I(\alpha) = \operatorname{sign}(1-\alpha)\int_0^{-\ln\alpha}\left\{\frac{1}{x+\alpha(1-e^x)} - \frac{1}{(1-\alpha)x}\right\}dx. \tag{2.3.49}$$

In Table 2.3.1 formula (2.3.48) is compared with numerical results for different values of ϵ and θ.

ϵ	θ=0.50		θ = 0.25		θ = 0.10	
	T_{as}	T_{num}	T_{as}	T_{num}	T_{as}	T_{num}
0.5	3.5359	4.6599	4.7247	5.1734	5.8567	6.0920
0.1	2.2470	2.3480	3.1303	3.1433	4.3014	4.3061
0.05	1.8668	1.8875	2.8015	2.8009	4.00945	4.00939
0.01	1.4320	1.4303	2.4612	2.4606	3.71766	3.71747
0.005	1.3557	1.3548	2.4058	2.4055	3.67143	3.67136
0.001	1.2816	1.2815	2.35364	2.35362	3.628628	3.628622

The period of solutions of (2.3.16) for different values of ϵ and θ. Values obtained from numerical integration are compared with the asymptotic expression (2.3.48).

Table 2.3.1

2.3.4 The period for large amplitude oscillations by inverse Laplace asymptotics

Starting point of the asymptotic investigation of Waldvogel (1983) is (2.3.1) with $e=0$, which after a scaling of the dependent variables, takes the form

$$\frac{dx}{dt} = ax(1-y), \tag{2.3.50a}$$

$$\frac{dy}{dt} = cy(-1+x). \tag{2.3.50b}$$

Then relation (2.3.4) representing the closed trajectories in the phase plane, becomes

$$\frac{1}{a}f(x)+\frac{1}{c}f(y)=H, \tag{2.3.51}$$

$$f(x)=x-1-\ln x.$$

We introduce the new variables

$$\xi=g(x),\quad \eta=g(y), \tag{2.3.52}$$

$$g(x)= \operatorname{sign}(x-1)\sqrt{2f(x)}$$

by which (2.3.51) is transformed into

$$\frac{1}{a}\xi^2+\frac{1}{c}\eta^2=2H. \tag{2.3.53}$$

A closed trajectory in the x,y-plane is mapped onto an ellipse with semi-axes

$$\alpha=\sqrt{2Ha},\ \beta=\sqrt{2Hc} \tag{2.3.54}$$

in the ξ,η-plane. The trajectories satisfy the differential equations

$$\frac{d\xi}{dt} = -\frac{a\eta G(\xi)G(\eta)}{G'(\xi)G'(\eta)} \tag{2.3.55a}$$

$$\frac{d\eta}{dt} = \frac{c\xi G(\xi)G(\eta)}{G'(\xi)G'(\eta)} \tag{2.3.55b}$$

where $G(u)$ is the inverse of $g(x)$ and $G'(u)$ is the derivative of $G(u)$ with respect to u. Hence the period $P(H)$ of a periodic solution is given by

$$P(H) = -\frac{1}{a}\int \frac{G'(\xi)G'(\eta)}{\eta G(\xi)G(\eta)}\,d\xi, \tag{2.3.56}$$

where the integral is taken clockwise over the ellipse in the ξ,η-plane. Introduction of the angular variable ϕ with

$$\xi=\alpha\cos\phi, \quad \eta=\beta\sin\phi \tag{2.3.57}$$

yields the expression

$$P(H)=\frac{2H}{\alpha\beta}\int_0^{2\pi} F(\alpha\cos\phi)\,F(\beta\sin\phi)\,d\phi, \tag{2.3.58}$$

with

$$F(u) = \frac{G'(u)}{G(u)}.$$

The Laplace transform of the formula for the period. The Laplace transform of $P(H)$, given by (2.3.58), is

$$p(s) = \frac{1}{\sqrt{ac}}\int_0^{\infty}\int_0^{2\pi} e^{-sH}F(\sqrt{2cH}\cos\phi)\,F(\sqrt{2aH}\sin\phi)\,d\phi dH. \tag{2.3.59}$$

The use of ξ and η, as intergration variables, separates the integration over the plane

$$p(s) = \frac{1}{ac}q(\frac{s}{a})q(\frac{s}{c}), \tag{2.3.60a}$$

$$q(s) = \int_{-\infty}^{\infty} e^{-1/2s\xi^2}F(\xi)d\xi. \tag{2.3.60b}$$

The integral (2.3.60b) can be expressed in the Gamma function as follows. We set $\xi=g(x)$, so that

$$q(s)=\int_0^{\infty} e^{-sf(x)}x^{-1}dx, \tag{2.3.61}$$

then, by taking $z=sx$ as integration variable, we obtain

$$q(s)=(\frac{s}{e})^{-s}\int_0^{\infty} e^{-z}z^{s-1}dz=(\frac{s}{e})^{-s}\Gamma(s), \tag{2.3.62}$$

$|\arg s| < \pi$.

Inverse Laplace asymptotics. The asymptotic behaviour of $P(H)$ for H large is found by inverse Laplace asymptotics. The integration path in the complex s-plane is taken counter clockwise along the negative real axis. Taking into account the singularities at $s=0$, $s=-ma$, $s=-mc$, $m=1, 2...$, we see that the main contribution comes from the singularity at $s=0$. Using the relation $\Gamma(s+1)=s\Gamma(s)$ and the series expansion for $\ln\Gamma(s+1)$ in terms of Riemann zeta functions (see appendix A), we obtain in view of the similarity rule of Laplace transformation

$$\frac{1}{A}p(\frac{s}{A})=\frac{A}{s^2}\exp\{-s(\gamma+\ln\frac{s}{\mu A})+ \qquad (2.3.63)$$

$$+\sum_{k=2}^{\infty}(z_k+\frac{\zeta(k)}{k})(-s)^k\},$$

$$A=1/a+1/c, \quad \mu=e(c^a a^c)^{1/(a+c)}, \qquad (2.3.64)$$

$$z_k=\frac{\zeta(k)}{k}\{\frac{(a^{-k}+c^{-k})}{A^k}-1\}.$$

where γ is the Euler constant and $\zeta(k)$ the Riemann zeta function. This expansion contains terms of the type

$$t_{km}(s)=s^k(\gamma+\ln s)^m. \qquad (2.3.65)$$

Their inverse Laplace transform is formally found from the inverse transform of the generating function

$$\chi(s,\nu)=e^{\gamma\nu}s^{k+\nu}=\sum_{m=0}^{\infty}\frac{1}{m!}\nu^m\, t_{km}(s), \qquad (2.3.66)$$

which is

$$\phi(H,\nu)=\frac{e^{\gamma\nu}}{H^{1+k+\nu}\Gamma(-\nu-k)}=\sum_{m=0}^{\infty}\frac{1}{m!}\nu^m T_{km}(H). \qquad (2.3.67)$$

Expanding the expression for $\phi(H,\nu)$ we obtain

$$\phi(H,\nu)=\frac{e^{-\nu\ln H}}{(-H)^{k+1}}\Gamma(1-\nu)\prod_{j=0}^{k}(j+\nu) \qquad (2.3.68)$$

or

$$\phi(H,\nu)=-\frac{1}{(-H)^{k+1}}\sum_{j=0}^{\infty}\frac{1}{j!}\nu^j(-\ln H)^j\sum_{m=0}^{\infty}\frac{1}{m!}\nu^m d_{km},$$

where

$$\sum_{m=0}^{\infty}d_{km}\frac{\nu^m}{m!}=\exp\{-\sum_{j=2}^{\infty}\frac{1}{j}\zeta(j)\nu^j\}\prod_{j=0}^{k}(j+\nu),$$

with

$$d_{k0}=0,\ d_{k1}=k!.$$

Thus, the inverse Laplace transform of $t_{km}(s)$ is

$$T_{km}(H)=\frac{1}{(-H)^{k+1}}\sum_{j=0}^{m}\begin{bmatrix} m \\ j \end{bmatrix} d_{kj}(-\ln H)^{m-j}. \tag{2.3.69}$$

Using (2.3.63), (2.3.65) and (2.3.69) we obtain

$$\frac{1}{A}P(AH)\approx H+L(H)+L(H)/H \tag{2.3.70}$$

$$-\{\frac{1}{2}L(H)^2-L(H)+z_2\}H^{-2}+\cdots$$

with

$$L(H)=\ln(\mu AH). \tag{2.3.71}$$

Waldvogel (1983) shows that it has computational advantages to consider the asymptotic series of the inverse. He finds for the first nine terms

$$\frac{1}{A}H(AP)\approx P-\ln(\mu AP)+\sum_{j=2}^{7}\frac{c_j}{P^j}, \tag{2.3.72}$$

where surprisingly c_j does not contain any logarithmic terms, see table 2.3.2.

	z_2	z_3	z_4	z_5	z_6	z_7	$\frac{1}{2}z_2^2$	z_2z_3	z_2z_4	z_2z_5	$\frac{1}{2}z_3^2$	z_3z_4	$\frac{1}{6}z_2^3$	$\frac{1}{2}z_2^2z_3$
c_2	1													
c_3	2	2												
c_4	3	9	6				6							
c_5	4	24	44	24			40	24						
c_6	5	50	175	250	120		150	240	120		120		120	
c_7	6	90	510	1350	1644	720	420	1260	1608	720	1620	720	1536	720

The coefficient c_i of formula (2.3.72). The table is read as follows: e.g. $c_4=3z_2+9z_3+6z_4+3z_2^2$.

Table 2.3.2

Exercises

2.3.1 Take specific values of ν and θ in (2.3.3) and integrate the system numerically.

2.3.2 Write the system (2.3.3) in the variables

$$p = \ln x \text{ and } q = \ln y.$$

Find a conservation function $C(p,q)$, such that

$$\frac{dp}{dt} = -\frac{\partial C}{\partial q}, \quad \frac{dq}{dt} = \frac{\partial C}{\partial p},$$

2.3.3 Show that small amplitude oscillations of (2.3.3) have a period of about $2\pi\sqrt{\nu}$.

2.4 CHEMICAL OSCILLATIONS

The study of biochemical oscillations began with the measurement of intracellular components in experiments on photosynthesis. Glycolytic oscillations were observed in a cell-free system of yeast. For more information on the biochemistry of this type of self-sustrained oscillations, we refer to Chance *et al.* (1973) and Goldbeter (1980). In this section we deal with a theoretical model of a chemical oscillation, the Brusselator, and an anorganic reaction scheme, the Belousov-Zhabotinskii reaction. For recent results on the dynamics of chemical oscillators the reader is refered to Rensing and Jaeger (1985).

2.4.1 The Brusselator

The following hypothetical chemical reaction was formulated by a Belgium group of physicists, see Nicolis and Prigogine (1977),

$$A \underset{k_{-1}}{\overset{k_1}{\rightleftarrows}} X, \tag{2.4.1a}$$

$$B + X \underset{k_{-2}}{\overset{k_2}{\rightleftarrows}} Y + D, \tag{2.4.1b}$$

$$2X + Y \underset{k_{-3}}{\overset{k_3}{\rightleftarrows}} 3X, \tag{2.4.1c}$$

$$X \underset{k_{-4}}{\overset{k_4}{\rightleftarrows}} E. \tag{2.4.1d}$$

Keeping the reactants A, B, D and E at a constant level and setting the reverse reactions all zero, we obtain for the concentrations of X and Y:

$$\frac{dx}{dt} = k_1 a - k_2 bx + k_3 yx^2 - k_4 x, \tag{2.4.2a}$$

$$\frac{dy}{dt}=k_2bx-k_3yx^2. \tag{2.4.2b}$$

Introduction of dimensionless variables defined by

$$\tau=k_4t,\quad u=k_4xy/(k_1a),\quad v=k_4/(k_1a), \tag{2.4.3}$$

$$\alpha=k_3(k_1a)^2/k_4^3,\quad \beta=k_2b/k_4 \tag{2.4.4}$$

transforms (2.4.2) into

$$\frac{du}{d\tau} = 1-(\beta+1)u+\alpha u^2v, \tag{2.4.5a}$$

$$\frac{dv}{d\tau} = \beta u-\alpha u^2v. \tag{2.4.5b}$$

This system has the equilibrium point $(\bar{u},\bar{v})=(1,\beta/\alpha)$, which is stable for $\alpha+1>\beta$. Varying β we find that above the critical value $\beta_c=1+\alpha$ a stable limit cycle with amplitude of order $O((\beta-\beta_c)^{1/2})$ branches off. This phenomenon is called a Hopf bifurcation see Appendix C. For $\beta>\alpha+1>>1$ with $\beta-\alpha=O(1)$ the oscillation turns into a relaxation oscillation. Introduction of the new dependent variable $w=u+v$ changes the system into

$$\frac{du}{d\tau} = (1-u)-\beta u+\alpha u^2(w-u), \tag{2.4.6a}$$

$$\frac{dw}{d\tau} = 1-u \tag{2.4.6b}$$

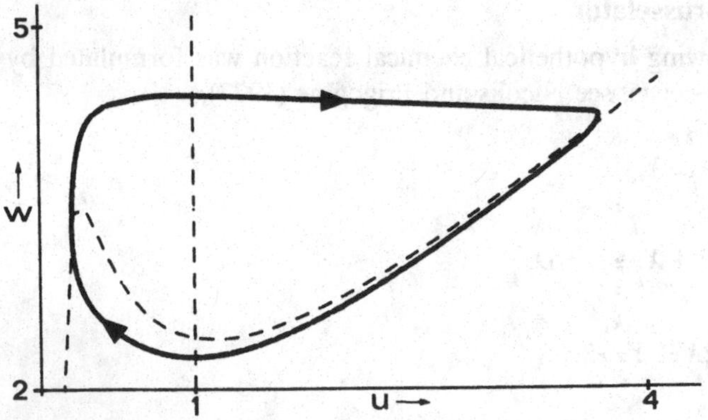

Fig.2.4.1 Relaxation oscillation of the Brusselator, see (2.4.6) with $\alpha=5$ and $\beta=7$. Most of the time the solution is near the two stable branches of $f=0$.

or

$$\frac{du}{d\tau} = f(u,w), \tag{2.4.6c}$$

$$\frac{dw}{d\tau} = g(u,w). \tag{2.4.6d}$$

A first inspection shows that the solution jumps from one stable branch to a second one and back similar to the Van der Pol relaxation oscillator, see fig.2.4.1. The branch near the w-axis approximately satisfies

$$w_1 = (\beta u - 1)/(\alpha u^2),$$

while for the other one we have

$$w_2 = u + \beta/(\alpha u).$$

The contribution to the period from the stable branches is

$$T_i = \int_{a_i}^{b_i} \frac{1}{1-u}\frac{dw_i}{du}du, \quad i = 1,2 \tag{2.4.7}$$

with

$$a_1 = \alpha/\beta^2, \; b_1 = 1/\beta, \; a_2 = \beta^2/(4\alpha) \text{ and } b_2 = \sqrt{\beta}/\alpha.$$

Consequently, we find for the period, after carrying out the integrations of (2.4.7),

$$T = \frac{\beta^2}{4\alpha} - \frac{\beta^3}{\alpha^2} + \frac{\beta^4}{\alpha^3} + 2\ln\beta - \ln\alpha + O(1). \tag{2.4.8}$$

2.4.2 The Belousov-Zhabotinskii reaction and the Oregonator

In 1958 Belousov described an oscillatory chemical reaction of anorganic type involving the oxidation of citric acid by potassium bromate. Later on Zhabotinskii modified some of the reactions and showed that certain components can be replaced by organic compounds. It is believed that biochemical and biological oscillations, which are much more complex and mostly not understood in detail, agree in many cases qualitatively with this Belousov-Zabotinskii reaction. For an extensive study of the reaction dynamics we refer to Tyson (1976). In this section we deal with a theoretical model of five irreversible reactions exhibiting about the same features as the Belousov-Zhabotinskii reaction. This model, known as the Oregonator, developed by Field and Noyes takes the form

$$A + Y \xrightarrow{k_1} X + P, \tag{2.4.9a}$$

$$X + Y \xrightarrow{k_2} 2P, \tag{2.4.9b}$$

$$A+X \xrightarrow{k_3} 2X+2Z, \tag{2.4.9c}$$

$$2X \xrightarrow{k_4} A+P, \tag{2.4.9d}$$

$$Z \xrightarrow{k_5} fY. \tag{2.4.9e}$$

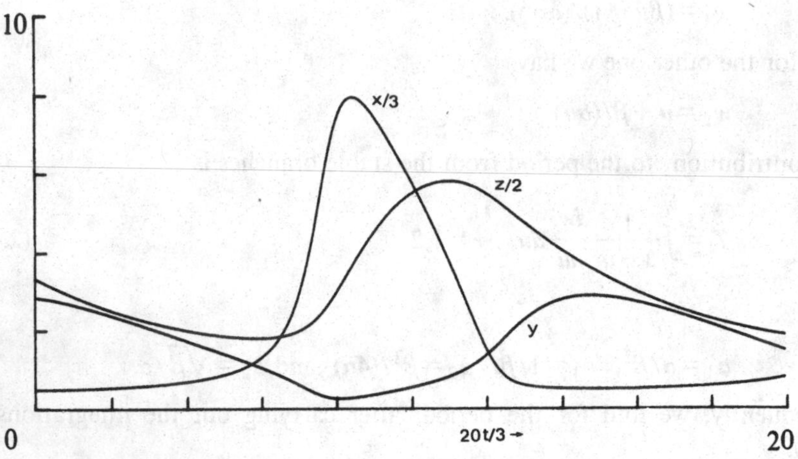

Fig.2.4.2 Periodic solution of the Field-Noyes equations (2.4.12) for $\epsilon=.2, f=.5, p=.161$ and $q=3.864\epsilon^3$.

Fixating the reactant A at a constant concentration a, we obtain for the concentrations of X, Y and Z a system of three coupled differential equations

$$\frac{dx}{dt} = k_1 ay - k_2 xy + k_3 ax - 2k_4 x^2, \tag{2.4.10a}$$

$$\frac{dy}{dt} = -k_1 ay - k_2 xy + f k_5 z, \tag{2.4.10b}$$

$$\frac{dz}{dt} = 2k_3 ax - k_5 z, \tag{2.4.10c}$$

where f is a suitable chosen stoichiometric coefficient. The following transformations are carried out to obtain dimensionless measures for the concentrations of X, Y and Z, representing bromous acid ($HBrO_2$), bromide ion (Br^-) and ceric ion (Ce^{+4})

$$u = k_2 x/(k_1 a), \quad v = k_2 y/(k_3 a), \tag{2.4.11a}$$

$$w = k_2 k_5 z/(2k_1 k_3 a^2), \tag{2.4.11b}$$

$$\tau = a\sqrt{k_1 k_3}\, t, \;\; p = k_5/(a\sqrt{k_1 k_3}), \tag{2.4.11c}$$

$$q = 2k_1 k_4/(k_2 k_3), \quad \epsilon = \sqrt{k_1/k_3}. \tag{2.4.11d}$$

We obtain

$$\epsilon\frac{du}{d\tau} = u + v - u\,v - q\,u^2, \tag{2.4.12a}$$

$$\frac{dv}{d\tau} = (2f\,w - v - uv)\,\epsilon\,, \tag{2.4.12b}$$

$$\frac{dw}{d\tau} = p\,(u - w). \tag{2.4.12c}$$

For certain values of p,q and f this system exhibits a relaxation oscillation see fig.2.4.2. A matched asymptotic solution has been given by Stanshine and Howard (1976), they assumed that $q = O(\epsilon^3)$. Nipp (1980) developed an algorithmic approach in his construction of appropriately scaled matched local solutions.

Exercises

2.4.1 Verify formula (2.4.8).

2.4.2 Take $q = \overline{q}\epsilon$ in (2.4.12). For which values of f, p and $\overline{q}$ does the system have a periodic solution. Derive the local asymptotic behaviour for at least two regions in the case, where the system exhibits a relaxation oscillation.

2.4.3 Formulate a reaction scheme that leads to the Volterra-Lotka equations (2.3.3), see Nicolis and Prigogine (1977).

2.5 BIFURCATION OF THE VAN DER POL EQUATION WITH A CONSTANT FORCING TERM

The Van der Pol equation with a constant forcing term exhibits a remarkable bifurcation when the stationary point at the unstable branch of $y = F(x)$, see (2.1.7), moves to the stable branch. The behaviour is in conflict with the idea we have about a Hopf bifurcation. In section 2.5.2 we solve this seeming contradiction. This forced Van der Pol equation is used in the modeling of excitability in biochemical systems, see section 2.5.1. The change of a continuous traffic flow into a run-stop cycle is a second application which is worth to be mentioned (Küne, 1984).

2.5.1 Modeling nerve excitation; the Bonhoeffer- Van der Pol equation

FitzHugh (1968) constructed a differential equation model explaining the phenomenon of "all or none" in nerve excitation. It means that a small stimulus results in a small disturbance of the stable equilibrium, while above a certain stimulus value (the threshold) the disturbance suddenly growths before the equilibruim state is restored. The model formulated by Hodgkin and Huxley in 1952 gives a precise description of the dynamics of nerve excitation as observed in the nerve cell of a squid. Their model consists of 4 coupled non-linear differential equations, in wich the variables represent physical quantities: the electric potential across the cell membrane and the concentrations of potassium, natrium and chloride ions. The question, posed by FitzHugh, is whether the phenomenon is also present in lower dimensional model equations. Such a metaphor for the threshold phenomenon are the FitzHugh-Nagumo equations, inwhich the dimension is brought down to 3. The Bonhoeffer- Van der Pol equation formulated by FitzHugh (1968) can be seen as an attempt to set the dimension at 2, see fig.2.5.1. Bonhoeffer (1948) gave a description in terms of isoclines for an iron wire model of the nerve cell. Fitzhugh made it explicit in the following set of equations

$$\epsilon \frac{dx}{dt} = y + x - \frac{1}{3}x^3 - s \tag{2.5.1a}$$

$$\frac{dy}{dt} = x - a + by, \quad 0 \leqslant b < 1, \quad a \geqslant 0, \tag{2.5.1b}$$

where s represents the stimulus being a positive constant over a short time interval. The variable x can be seen as the electric potential across the cell membrane, y does not have any physical meaning.

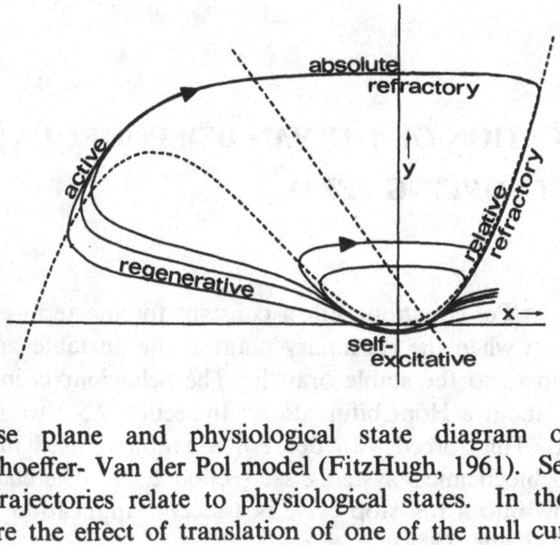

Fig.2.5.1 Phase plane and physiological state diagram of the Bonhoeffer- Van der Pol model (FitzHugh, 1961). Sections of trajectories relate to physiological states. In the next figure the effect of translation of one of the null curves is shown.

For $a<1-2/3b$ the system has a stable equilibrium $(\bar{x},\bar{y})$ with $\bar{x}>1$. For $s>s_c$ with

$$s_c=\frac{2}{3}+\frac{a-1}{b} \tag{2.5.2}$$

this equilibrium looses its stability. In fig.2.5.2 we sketch the dynamics of the system with ϵ small and a stimulus $s>s_c$ for $0<t<\delta$ ($s=0$ elsewhere) Recently, Kopell and Ermentrout (1985) have analyzed a variant of (2.5.1) with the second null curve having a *s*-shape instead of a line.

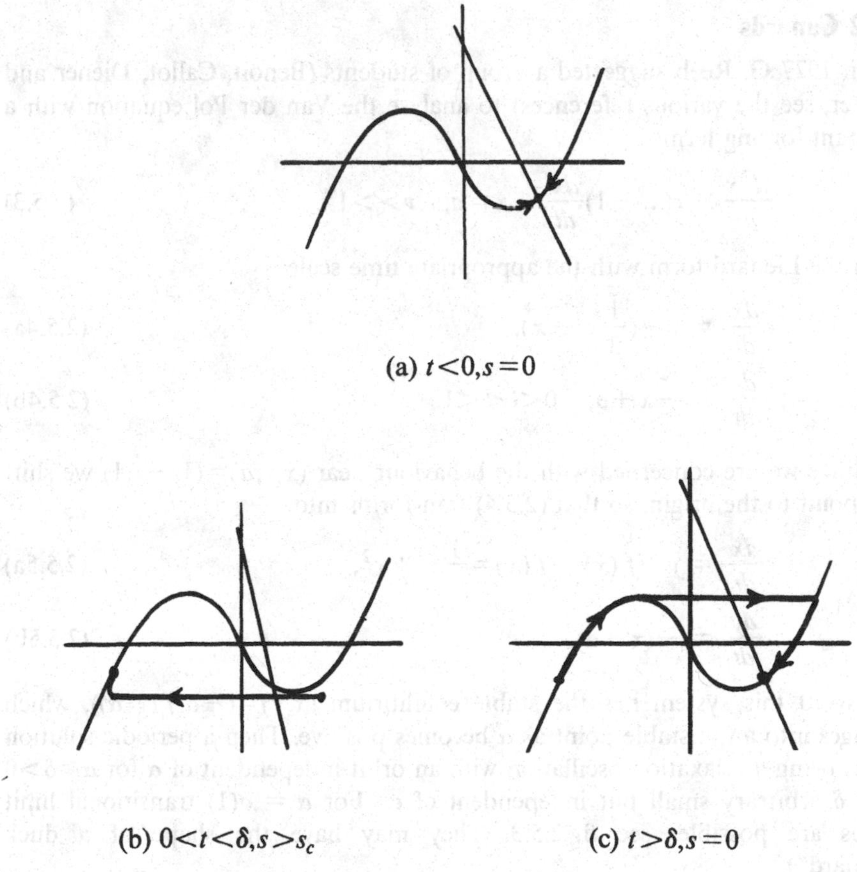

(a) $t<0, s=0$

(b) $0<t<\delta, s>s_c$ (c) $t>\delta, s=0$

Fig.2.5.2 The Bonhoeffer- Van der Pol model for $\epsilon=0$. For $s=0$ the system has a stable equilibrium at a stable branch, while for $s>s_c$ the system describes a relaxation oscillation.

A similar phenomenon occurs in the adenylate cyclase reaction and other excitable biochemical systems, see Goldbeter (1980) and Hahn *et al.* (1974).

We see that, as the parameter s crosses the value s_c, a Hopf bifurcation takes place. That is, the stable equilibrium becomes unstable and a periodic solution branches off. However, the present situation seems different from the usual picture as the oscillation has right away a large amplitude. In order to obtain a better understanding of what is going on, s has to be stretched at s_c with ϵ as small parameter. Instead of carrying out this analysis we reduce the problem somewhat. Examining the null curves and their dependence upon the parameter, it is seen that the same occurs if we set $s=b=0$ and let a cross the value 1. This reduced problem will be discussed in the next section.

2.5.2 Canards

In 1977 G. Reeb suggested a group of students (Benoit, Callot, Diener and Diener, see the various references) to analyse the Van der Pol equation with a constant forcing term

$$\frac{d^2x}{dt^2}+\nu(x^2-1)\frac{dx}{dt}+x=a, \quad \nu>>1 \tag{2.5.3}$$

or in the Lienard form with the appropriate time scale

$$\epsilon\frac{dx}{dt}=y-(\frac{1}{3}x^3-x), \tag{2.5.4a}$$

$$\frac{dy}{dt}=-x+a, \quad 0<\epsilon<<1. \tag{2.5.4b}$$

Since we are concerned with the behaviour near $(x,y,a)=(1,-1,1)$ we shift this point to the origin, so that (2.5.4) transforms into

$$\epsilon\frac{dx}{dt}=y-f(x), \quad f(x)=\frac{1}{3}x^3+x^2, \tag{2.5.5a}$$

$$\frac{dy}{dt}=-(x+\alpha). \tag{2.5.5b}$$

For $\alpha<0$ this system has the stable equilibrium $(\overline{x},\overline{y})=(-\alpha,f(-\alpha))$, which changes into an unstable point as α becomes positive. Then a periodic solution arises being a relaxation oscillation with an orbit independent of α for $\alpha\geqslant\delta>0$ with δ arbitrary small but independent of ϵ. For $\alpha=o(1)$ transitional limit cycles are possible, see fig.2.5.3. They may have the shape of a duck ("canard").

Instead of explaining the method of analysis developed by the group, for whom this problem served as an example of the application of nonstandard analysis, we follow the asymptotic method of Eckhaus (1983) which is more standard for our approach of asymptotics for relaxation oscillations.

Let $\alpha(\epsilon)=O(1)$. A trajectory near the stable branch CD is studied by making the following local transformation

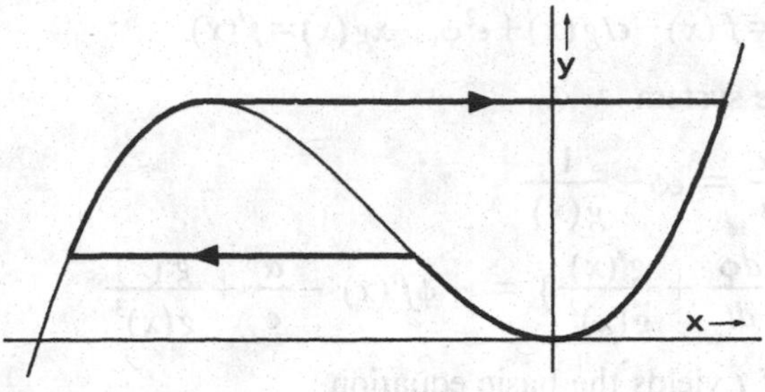

(a) the canard. Smooth transitions to a regular relaxation oscillation and to (b) are possible.

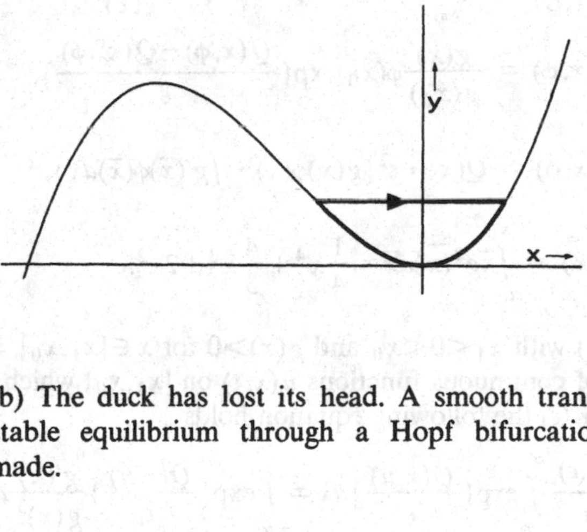

(b) The duck has lost its head. A smooth transition to a stable equilibrium through a Hopf bifurcation can be made.

Fig.2.5.3 Limit cycles of the system (2.5.5) with $\alpha = o(1)$

$$y=f(x)-\epsilon/g(x)+\epsilon^2\phi, \quad xg(x)=f'(x), \tag{2.5.6}$$

so we have the system

$$\frac{dx}{dt} = \epsilon\phi-\frac{1}{g(x)}, \tag{2.5.7a}$$

$$\epsilon\{\frac{d\phi}{dt}+\frac{g'(x)}{g(x)^2}\} = -\phi f'(x) - \frac{\alpha}{\epsilon} + \frac{g'(x)}{g(x)^3}. \tag{2.5.7b}$$

Elimination of t yields the basic equation

$$\epsilon\{-(\frac{\phi}{g(x)})'+\epsilon\phi\phi'\}=-f'-\frac{\alpha}{\epsilon}+\frac{g'}{g(x)^3}. \tag{2.5.8}$$

For this equation of ϕ we can formulate an equivalent problem

$$\phi = \chi_1(x,\phi)+\chi_2(x,\phi), \tag{2.5.9a}$$

$$\chi_1(x,\phi) = \frac{g(x)}{\epsilon}\exp\{\frac{Q(x,\phi)}{\epsilon}\} \tag{2.5.9b}$$

$$\times\int_{x_0}^{x}\exp\{-\frac{Q(\bar{x},\phi)}{\epsilon}\}\{\frac{\alpha}{\epsilon}-\frac{g'(\bar{x})}{g(\bar{x})^3}\}d\bar{x},$$

$$\chi_2(x,\phi) = \frac{g(x)}{g(x_0)}\phi(x_0)\exp\{\frac{Q(x,\phi)-Q(x_o,\phi)}{\epsilon}\}, \tag{2.5.9c}$$

$$Q(x,\phi) = \hat{Q}(x)+\epsilon^2\{g(x)\phi(x)-\int_0^x g'(\bar{x})\phi(\bar{x})d\bar{x}\}, \tag{2.5.9d}$$

$$\hat{\hat{Q}}(x) = \int_0^x \bar{x}g^2(\bar{x})d\bar{x}=\frac{1}{4}x^4+\frac{4}{3}x^3+2x^2. \tag{2.5.9e}$$

Let $x\in(x_1,x_0)$ with $x_1<0<x_0$ and $g(x)>0$ for $x\in[x_1,x_0]$, and let moreover U be the set of continuous functions $u(x,\epsilon)$ on $[x_1,x_0]$ which are bounded for $\epsilon\to0$. For $\alpha=\alpha_c(\epsilon)$ the following equation holds

$$\frac{\alpha_c(\epsilon)}{\epsilon}\int_{-\infty}^{\infty}\exp\{\frac{Q(x,u)}{\epsilon}\}dx=\int_{-\infty}^{\infty}\exp\{\frac{Q(x,u)}{\epsilon}\}\frac{g'(x)}{g(x)^3}dx \tag{2.5.10}$$

with $u=0$. In this equation the functions $g(x)$ and $u(x)$ have arbitiary continuous prolongations outside the interval (x_1,x_0) which change the value of $\alpha_c(\epsilon)$ by an exponentially small contribution. A similar change would be found if instead of $u=0$ any $u\in U$ were substituted. In a first order approximation we have

$$\alpha_c(\epsilon)=\epsilon g'(0)/g(0)^3+O(\epsilon^2). \tag{2.5.11}$$

Thus, if α is exponentially close to α_c, we have that

$$\chi_1(x,u)=\chi_1(x,0)+O(\epsilon). \tag{2.5.12}$$

Moreover,if x_0 is sufficiently large $Q(x)-Q(x_0)$ is negative for $x\in(x_1,x_0)$

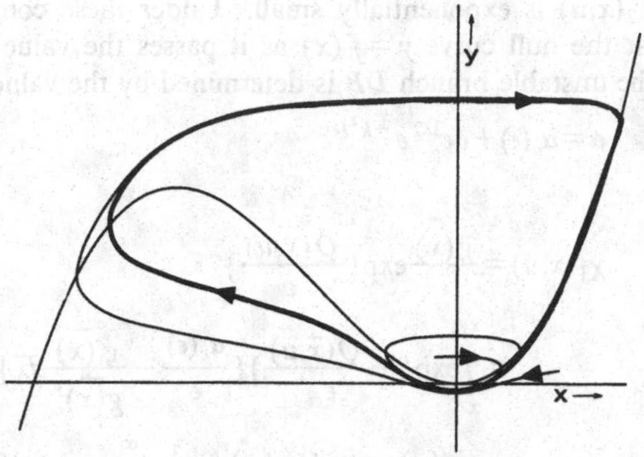

(a) the duck

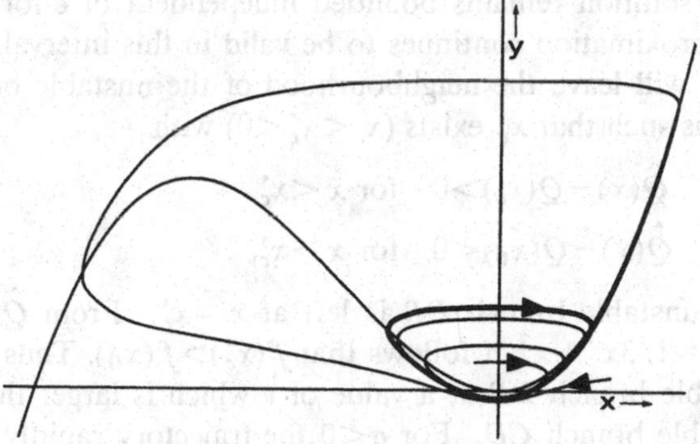

(b) the duck without a head

Fig.2.5.4 Trajectories of (2.5.5) that approach the duck with(out) head and so $\chi_2(x,u)$ is exponentially small. Under these conditions the trajectory stays near the null curve $y=f(x)$ as it passes the value $x=0$. The point of leaving the unstable branch DB is determined by the value of α. Let

$$\alpha=\alpha_c(\epsilon)+\sigma\epsilon^{3/2}e^{-k^2/\epsilon}, \tag{2.5.13}$$

then

$$\chi_1(x,u)=\frac{g(x)}{\epsilon}\exp\{\frac{Q(x,u)}{\epsilon}\} \tag{2.5.14}$$

$$\times\int_x^{x_1}\exp\{\frac{-Q(\bar{x},u)}{\epsilon}\}\{\frac{\alpha_c(\epsilon)}{\epsilon}-\frac{g'(\bar{x})}{g(\bar{x})^3}d\bar{x}\}+$$

$$-\frac{\sigma g(x)}{\sqrt{\epsilon}}\exp\{\frac{Q(x,u)-k^2}{\epsilon}\}\int_{-\infty}^{\infty}\exp\{-\frac{Q(\bar{x},u)}{\epsilon}\}d\bar{x}.$$

We assume that k is such that for some $x_c\in(-2,0)$ the following equation holds

$$Q(x_c)-k^2=0. \tag{2.5.15}$$

Then the solution remains bounded independent of ϵ for $x_c\in(-2,0)$ and the linear approximation continues to be valid in this interval. At $x=x_c+O(\epsilon)$ the trajectory will leave the neighbourhood of the unstable branch. If the starting point x_0 is such that x_c^* exists $(x_c<x_c^*<0)$ with

$$\hat{Q}(x)-Q(x_o)>0 \quad \text{for } x<x_c^*, \tag{2.5.16a}$$

$$\hat{Q}(x)-Q(x_0)<0 \quad \text{for } x>x_c^*, \tag{2.5.16b}$$

then the unstable branch DB is left at $x=x_c^*$. From $\hat{Q}(x)$ given by (2.5.9e) and $f(x)=1/3x^3+x^2$ it follows that $f(x_c^*)>f(x_0)$. Thus the trajectories leave the unstable branch DB at a value of y which is larger than the starting value at the stable branch CD. For $\sigma<0$ the trajectory rapidly returns to the stable branch CD. Repeating the procedure of following the trajectory as it passes along the point D, we see that it spirals outwards and that x_c^* converges to x_c. For $\sigma>0$ it jumps at x_c^* to the stable branch AB. In the next cycle (and all the following ones) it jumps at x_c, see fig.2.5.4. For increasing k, the coordinate x_c approaches the value -2. At this value the two limit cycles ($\sigma>0$ and $\sigma<0$) merge.

Eckhaus (1983) also analyses the case $g'(0)<0$, which has a more complicated pattern of bifurcating periodic solutions.

Exercises

2.5.1 Compute the eigenvalues of the linearized system in the equilibrium of (2.5.1) and analyse the dependence upon s.

2.5.2 Construct two coupled Bonhoeffer-Van der Pol equations, such that one of them stimulates the other.

2.5.3 Construct the asymptotic solution of (2.5.5) with $\alpha=\alpha_c(\epsilon)$ for the local region at $(x,y) = (-2,4/3)$.

2.5.4 Analyse the asymptotic solution of

$$\epsilon\frac{dx}{dt} = y - \frac{1}{2}x^2,$$

$$\frac{dy}{dt} = -x - a\sqrt{\epsilon}$$

for different values of a. Find also a transformation that eliminates ϵ (Lauwerier, 1985).

2.5.5 Analyse the Hopf bifurcation of the Van der Pol equation

$$\frac{d^2x}{dt^2} + \nu(x^2 - a)\frac{dx}{dt} + x = 0.$$

2.6 STOCHASTIC AND CHAOTIC OSCILLATIONS

Physical systems with periodic behaviour usually show fluctuations in the period. One is inclined to ascribe this to perturbations coming from the environment (noise). The possibility of chaotic behaviour of deterministic differential equation systems with three or more state variables has recently turned the attention to an explanation in this direction. In section 2.4.2, we modeled the Belousov-Zhabotinskii reaction by a system of three differential equations having a periodic solution in a certain parameter range. Rössler and Wegman (1978) and Lozi (1982) have shown that a slightly different model of this reaction may produce chaotic behaviour.

Chaotic solutions of a system of coupled nonlinear differential equations have been noticed the first time in the Lorenz equations, see Lorenz (1963) and Guckenheimer and Holmes (1984). A different way to study chaotic dynamics is to consider iterates of a mapping of a compact domain $\mathbb{R}^n$ into itself. Trajectories of a dynamical system that remain in a bounded domain of state space and repeatedly return in a transversal intersection Σ induce such a mapping of Σ into itself (the Poincaré mapping). A fixed point of the mapping or a finite repetition of the mapping corresponds with a periodic solution, see also appendix C.

The simplest mapping showing periodic as well as chaotic behaviour is the logistic map

$$f{:}x \mapsto ax(1-x), \quad 0<a\leqslant 4 \tag{2.6.1}$$

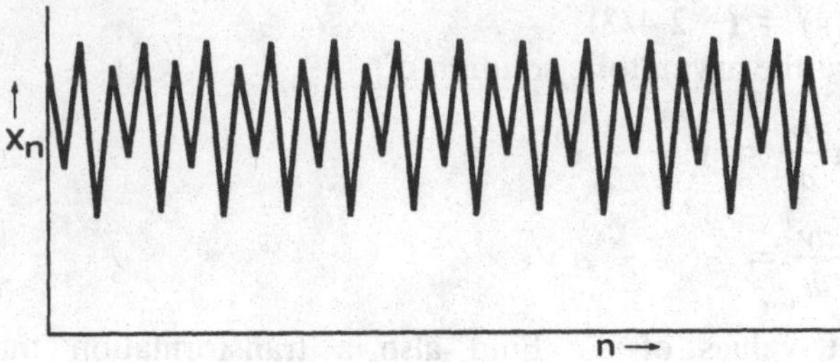

(a) period 8 solution with $a = 3.55$ and $\delta = 0$

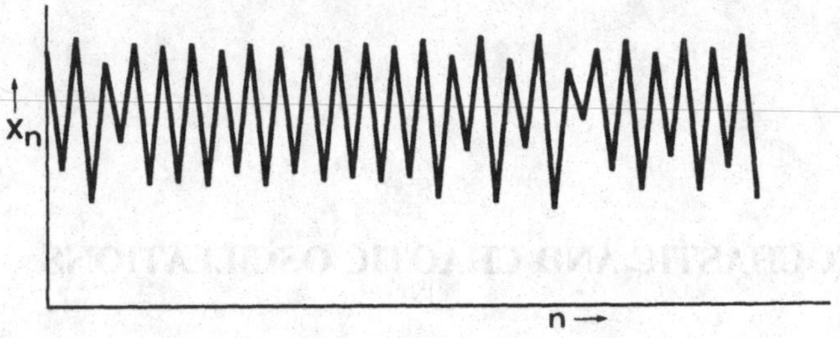

(b) chaotic solution with $a = 3.60$ and $\delta = 0$

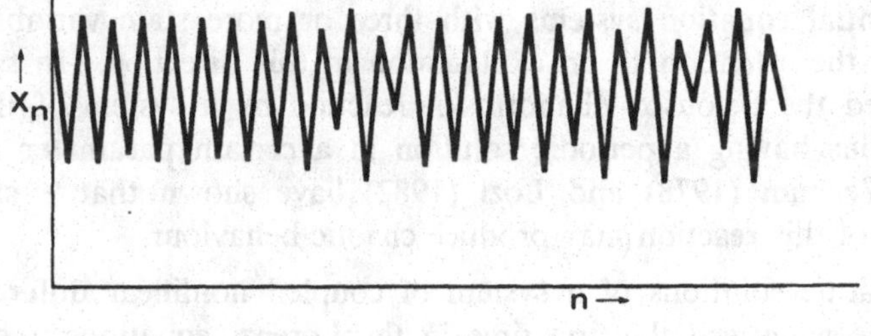

(c) period 8 solution with $a = 3.55$ and $\delta = 0.025$

Fig.2.6.1 The logistic difference equation with and without random perturbations. It is noted that the iterates (b) and (c) look similar although the underlying mechanism causing random behaviour is different.

for $x \in (0,1)$, see May (1967) and Collet and Eckmann (1980). For $0<a<3$ the sequence of points, found by repeated mapping, behaves quite regular. It converges independently from the starting value to either $0\,(a \leqslant 1)$ or to $1-1/a$ $(a>1)$. For $a>3$ periodic limit behaviour is found. A sequence of values $a_m, m=1,2,3,...$ exists, such that for $a \in (a_m, a_{m+1})$ the periodic limit set consists of 2^m points (repeated period doubling),

$$a_1 = 3, \quad a_2 = 3.449, a_3 = 3.544 \quad \text{and} \quad a_\infty = 3.570.$$

In fig.2.6.1 we give a period 8 solution ($a=3.55$), a chaotic solution ($a=3.60$) a period 8 solution ($a=3.55$) with random perturbations satisfying

$$X_{n+1} = aX_n(1-X_n) + \delta g_n, \tag{2.6.2}$$

with g_n a random number with a homogeneous distribution over the interval $[-0.5,0.5]$.

Under investigation is the problem of detecting chaotic solutions from the output of a dynamical system perturbed by noise, see the analysis of section 2.6.4. A way to explore the distinction between stochastic and chaotic behaviour is to compare a stochasticly perturbed two component model having a periodic solution with a three component model having a chaotic solution. In a chemical system such a third component can be an intermediate reactant, which is present for a short time and therefore not noticed (hidden), see Engel-Herbert *et al.* (1985). On the other hand noise may regularize chaotic dynamics and make it look periodic, see Matsumoto and Tsuda (1983) and Matsumoto (1984). They speak of hidden (chaotic) dynamics.

In the next section we deal with chaotic relaxations oscillations; we explain their structure and give, as an example, a model for irregular electric activity of nerve cells. Moreover, a chaotic Van der Pol oscillator with a hidden variable is constructed. In sections 2.6.2 and 2.6.3 stochasticly perturbed relaxations are analyzed and for the Van der Pol oscillator the distribution of the period is computed.

2.6.1 Chaotic relaxation oscillations

From studies of Lorenz (1963), Smale (1967) and Ruelle and Takens (1971) it is known that dynamical systems with at least 3 state variables may exhibit chaotic behaviour. It is quite well possible to construct a system of type (2.1.5) possessing a limit solution being a strange atractor and having the properties (2.1.6).

By example we show that a system, that remains most of the time in a 2-dimensional manifold, still may exhibit chaotic dynamics. Let us consider a system with one fast and two slow variables of the type depicted in 1.4.1. In fig.2.6.2, trajectories are sketched. In the discontinuous limit approximation there is a closed trajectory through A. In fig.2.6.2 it is seen that the Poincaré mapping of the compact interval AB into itself agrees qualitatively with the logistic map, see (2.6.1). The phenomena of period doubling and chaos are present in this mapping. Systematic studies on the occurrence of chaos in con-

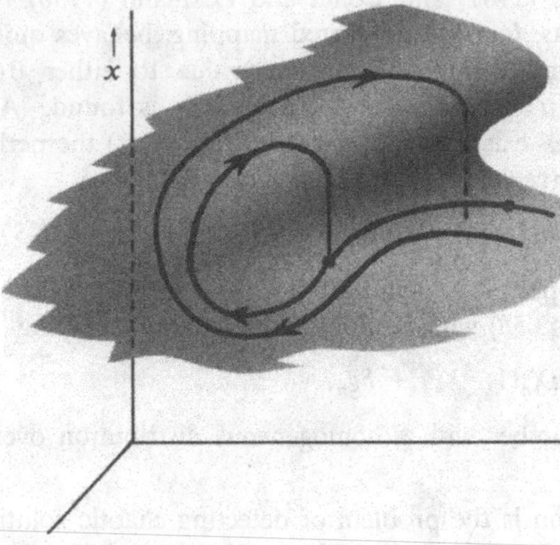

(a) discontinuous solutions

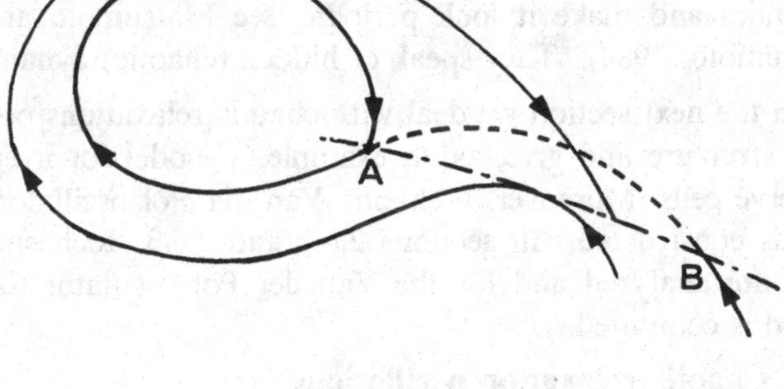

(b) trajectories projected in the y-plane

Fig.2.6.2 Example of a relaxation oscillator with chaotic dynamics. Following one cycle the trajectories generate a mapping of the curve AB into itself.

strained equations are carried out by Oka and Kokubu (1985ab) and Takens (1986).

Chaotic relaxation oscillations are met in the modeling of electric activity of nerve cells treated with barium, see Argémi and Rossetto (1983). The metaphorical model of the biological system is of the form

$$\epsilon\frac{dx}{dt} = -x^3 + y_1 x - y_2, \quad \epsilon = 0.05,$$

$$\frac{dy_1}{dt} = 1 - 0.1x - y_1,$$

$$\frac{dy_2}{dt} = (1 - 0.1x - y_1)x + x + \lambda$$

and exhibits its first period doubling at $\lambda = -0.555916$. Chaotic solutions are found near $\lambda = -0.555846$. More examples with applications in fluid mechanics and biology are given by Lozi (1983). Ito *et al.* (1980) model the timing of earthquakes by a chaotic relaxation oscillator.

To the Van der Pol oscillator a third state variable can be added, such that the system of equations for the three variables exhibits chaotic dynamics, while the regular Van der Pol relaxation oscillation is found if the third variable is set zero (Grasman and Roerdink; 1986). This is achieved by construction of a vector field similar to the one of fig. 2.6.2. We consider the system

$$\epsilon\frac{dx}{dt} = y_1 - \frac{1}{3}x^3 + x, \tag{2.6.3a}$$

$$\frac{dy_1}{dt} = -x - x^2 y_2, \tag{2.6.3b}$$

$$\frac{dy_2}{dt} = (x + a)y_2^2. \tag{2.6.3c}$$

In fig. 2.6.3a we sketch the trajectories in the x,y_2-plane. In the limit $\epsilon \to 0$ the solution jumps from $x=1$ to $x=-2$ and from $x=-1$ to $x=2$. The Poincaré mapping of the line

$$(x,y_1,y_2) = (2,-2/3,y_2)$$

into intself is given in fig. 2.6.3b for the case $a=1.7$. For $a \in (1.0,\ 2.0)$ repeated period doubling and chaos are found.

In fig. 2.6.3c a chaotic solution of (2.6.3) with $a=1.7$ and $\epsilon=.05$ is given. The Lyapunov exponents of this chaotic solution are

$$\lambda_1 = 0.09, \ \lambda_2 = 0 \quad \text{and} \quad \lambda_3 = -26.45,$$

see Wolf *et al.* (1985) for the definition and computation of these exponents. A positive Lyapunov exponent indicates chaotic behaviour.

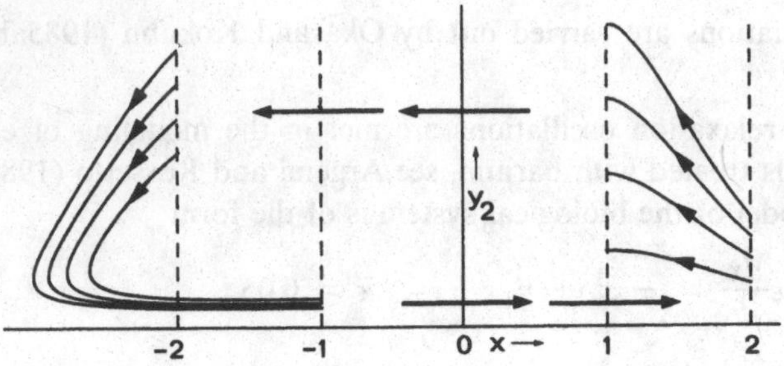

(a) trajectories projected in the x,y_2-plane

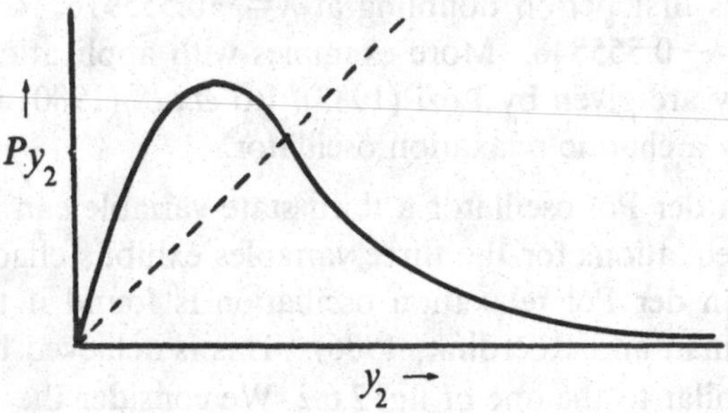

(b) the Poincaré mapping at $(x,y_1)=(2,-2/3)$

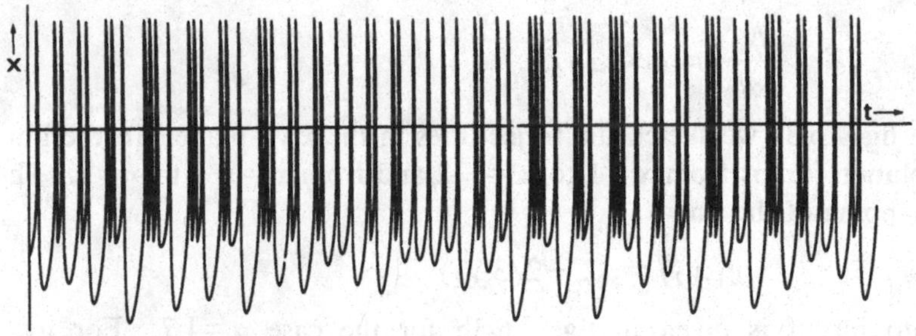

(c) a chaotic solution of (2.6.3) with $(\epsilon,a)=(0.05,1.7)$

Fig.2.6.3 A chaotic Van der Pol type relaxation oscillator for $\epsilon\rightarrow 0$. For $y_2=0$ it is identical to the Van der Pol oscillator, see formula (2.6.3).

2.6.2 Randomly perturbed oscillations

In this section we consider the influence of random perturbations upon the generalized Van der Pol oscillator (2 .8.3). For an application in biochemistry we refer to Hahn *et al.* (1974). Ebeling *et al.* (1986) analyse stochastic dynamics of the Duffing -Van der Pol oscillator.

Since after perturbation the fast variables x_i rapidly return to the stable equillibrium of (2.1.83a) with y constant, we only need to take into consideration perturbations of the equation for y. Thus, we analyse the system of stochastic differential equations

$$\epsilon dX_i = F_i(X,Y)dt, \quad i = 1,...,m,$$

$$dY_j = G_j(X,Y)dt + \delta \sum_{k=1}^{p} \sigma_{jk}(X,Y)dW_k, \quad j = 1,...,n.$$

It is assumed that in (2.6.2) $0<\epsilon<<\delta<<1$.

In our perturbation analysis we let $\epsilon \to 0$ and consider the discontinuous periodic solution L_0 of the deterministic system and its stochastic perturbations $W_1,..., W_p$ denoting p independent Wiener processes. We will apply methods for stochastic differential equations as described in the book of Gardiner (1983).

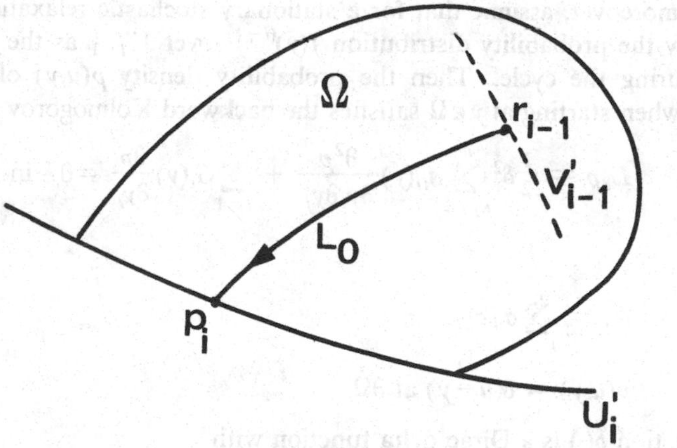

Fig.2.6.4 The section of the discontinuous limit cycle L_0 that lies between r_{i-1} and p_i is projected in the y-space. The set of return points V_{i-1} near r_{i-1} and jump points U_i near p_i are also projected. All stochastic trajectories leave Ω through U'_i with probability 1 for a sufficiently large domain Ω.

In the study on the existence of periodic solutions of the generalized Van der Pol equation (section 2.1.4), we defined the set of points U_i in a neighbourhood of a singular point p_i on the discontinuous closed trajectory L_0:

$$U_i = \{(x,y)|F(x,y) = 0, \det A(x,y) = 0\}$$

with A given by (2.1.84). The set of points V_i in a neighbourhood of the return point r_i of L_0 is defined by

$$V_i = \{(x,y)|F(x,y) = 0, (\bar{x},y) \in U_i \text{ for some } \bar{x}\}. \tag{2.6.4}$$

On the stable manifold $\mathcal{F}$ stochastic trajectories of the reduced system ($\epsilon=0$) connect points of V_{i-1} with points of U_i and satisfy

$$dY_j = G_j(\chi(Y),Y)dt + \delta \sum_{k=1}^{p} \sigma_{jk}(\chi(Y),Y)dW_k, \tag{2.6.5}$$

where $x = \chi(y)$ is a local solution of $F(x,y)=0$. In y-space we analyse stochastic trajectories in a domain Ω being a sufficiently large neighbourhood of the projection of the segment of L_0 that connects r_{i-1} with p_i. At one side this domain is bounded by U'_i, being the projection of U_i in the y-space, see fig.2.6.4. The boundary U_i' is an absorbing boundary for (2.6.5). The remaining part of $\partial\Omega$ is at large distance of L'_0, so that it does not play any role in the stochastic analysis.

Let at some time a stochastic trajectory arrive in a point with $y=u \in U_i'$. Let us, moreover, assume that for a stationary stochastic relaxation oscillation, we know the probability distribution $f(y)^{(i-1)}$ over V'_{i-1} as the system passes V_{i-1} during the cycle. Then the probability density $p(u,y)$ of exit through $u \in U_i'$, when starting at $y \in \Omega$ satisfies the backward Kolmogorov equation

$$L_\delta p \equiv \frac{1}{2}\delta^2 \sum_{i,j=1}^{n} a_{ij}(y)\frac{\partial^2 p}{\partial y_i \partial y_j} + \sum_{i=1}^{n} G_i(y)\frac{\partial p}{\partial y_i} = 0 \quad \text{in} \quad \Omega, \tag{2.6.6}$$

where

$$a_{ij} = \sum_{k=1}^{p} \sigma_{ik}\sigma_{kj},$$

$$p(u,y) = \delta(u-y) \text{ at } \partial\Omega.$$

The function $\delta(\cdot)$ is a Dirac delta function with

$$\int \delta(u-y)h(y)dS_{U_i} = h(u) \quad \text{for} \quad u \in \partial\Omega.$$

Consequently, if the distribution of starting points at $y \in V'_{i-1}$ is $f^{(i-1)}(y)$, then

$$f^{(i)}(u) = \int_{V_{i-1}} p(u,y) f^{(i-1)}(y) dS_{V_{i-1}} \tag{2.6.7}$$

is the distribution of arrival points at U_i'.

Continuing through the cycle, we obtain as many of these equations as there are jumps. From these equations we may determine $f^{(i)}(y)$. Since the deterministic trajectory L_0 is stable, stochastic trajectories remain most likely in a δ-neighbourhood of L_0 and we may make suitable approximations. Diffusion parallel to the deterministic flow is neglected and, moreover, the distribution at U_i' is approximated by

$$f^{(i)}(y) = \frac{k_i}{\delta^{n-1}}\exp\{-\frac{(y-y_{r_i})^T A_i (y-y_{r_i})}{\delta^2}\}, \tag{2.6.8}$$

Then a system of equations for A_i can be formulated.

Let for a point $y \in V'_i$ the time $T_i(y)$ needed to reach U'_{i+1} be a stochastic variable with distribution $g_i(t,y)$. Then the first and second moment, defined as

$$\hat{T}_i(y) = \int_0^\infty t g_i(t,y)dt, \quad \hat{S}_i(y) = \int_0^\infty t^2 g_i(t,y)dt \tag{2.6.9}$$

satisfy

$$L_\delta \hat{T}_i = -1 \text{ in } \Omega, \tag{2.6.10a}$$

$$\hat{T}_i = 0 \text{ on } U'_{i+1} \tag{2.6.10b}$$

$$L_\delta \hat{S}_i = -2\hat{T}(y) \text{ in } \Omega, \tag{2.6.11a}$$

$$\hat{S}_i = 0 \text{ on } U'_{i+1}. \tag{2.6.11b}$$

The distribution of time between jump i and $i+1$ is

$$h_i(t) = \int_{V_{i'}} g_i(t,y) f^{(i)}(y) dS_{V_i}. \tag{2.6.12}$$

Its first and second moment are

$$\hat{T}_i(V_i) = \int_{V_i'} \hat{T}_i(y) f^{(i)}(y) dS_{V_i}, \tag{2.6.13}$$

$$\hat{S}_i(V_i) = \int_{V_i'} \hat{S}_i(y) f^{(i)}(y) dS_{V_i}. \tag{2.6.14}$$

2.6.3 The Van der Pol oscillator with a random forcing term

For the Van der Pol oscillator the set U_i is just one point and, because of symmetry, we only have to compute one distribution of time between two jumps. We take the following stochastic forcing

$$\epsilon dX = (Y - \frac{1}{3}X^3 + X)\,dt \tag{2.6.15a}$$

$$dY = -Xdt + \delta dW. \tag{2.6.15b}$$

For $\epsilon \to 0$ the solution satisfies $X = X_\pm(Y)$, which are roots of the equation

$$Y = \frac{1}{3}X^3 - X \tag{2.6.16}$$

for $X>1$ and $X<-1$. Let us analyse the stochastic trajectories on the negative branch. The stochastic differential equation for the reduced problem ($\epsilon=0$) reads

$$dY = -X_-(Y)dt + \delta dW, \tag{2.6.17a}$$

$$Y(0) = y, \quad y<2/3. \tag{2.6.17b}$$

The domain is bounded by a reflecting boundary for $y\to-\infty$ and an absorbing boundary at $y=2/3$. Let the stochastic exit time $T(y)$ have a distribution $f(t,y)$. The first and second moment of this distribution satisfy the following elliptic boundary value problems

$$L_\delta \hat{T} = -1, \quad \hat{T}_y(-\infty) = 0, \quad \hat{T}(2/3) = 0 \tag{2.6.18}$$

and

$$L_\delta \hat{S} = -2\hat{T}(y), \quad \hat{S}_y(-\infty) = 0, \quad \hat{S}(2/3) = 0, \tag{2.6.19}$$

where L_δ is given by (2.6.6) with $a_{11}=1$ and $G_1(y)=-X_-(y)$. Integral expressions for the solutions are

$$\hat{T}(y,\delta) = \frac{2}{\delta^2} \int_y^{2/3} \int_{-\infty}^{u} e^{2\{R(u)-R(z)\}/\delta^2} dzdu, \tag{2.6.20}$$

and

$$\hat{S}(y,\delta) = \frac{4}{\delta^2} \int_y^{2/3} \int_{-\infty}^{u} \hat{T}(z;\delta) e^{2\{R(u)-R(z)\}/\delta^2} dzdu, \tag{2.6.21}$$

where

$$R(y) = \int_{-\infty}^{y} X_-(u)du. \tag{2.6.22}$$

The integrals can be evaluated asymptotically for $0<\delta<<1$, yielding

$$\hat{T}(y,\delta) = \hat{T}_0(y) + \delta^2 \hat{T}_1(y) + \cdots, \tag{2.6.23}$$

with

$$\hat{T}_0 = -\frac{1}{2} + \frac{1}{2}X_-^2 - \ln(-X_-),$$

$$\hat{T}_1 = \frac{1}{4}\left(\frac{1}{X_-^2} - 1\right).$$

It is noted that $T_0(y)$ is the time the reduced deterministic trajectory ($\epsilon=\delta=0$), starting at y, needs to reach the value 2/3. Likewise we have

$$\hat{S}(y;\delta) = \hat{S}_0(y) + \delta^2 \hat{S}_1(y) + \cdots, \tag{2.6.24}$$

with

$$\hat{S}_0 = \hat{T}_0^2,$$

$$\hat{S}_1 = -\frac{1}{4} - \frac{1}{4}X_-^2 + \frac{1}{2X_-^2} + 2\ln(-X_-).$$

Consequently, the time between two jumps has a distribution with the following expectation and variance:

$$E(T) = \hat{T}(-2/3;\delta), \tag{2.6.25a}$$

$$\mathrm{Var}(T) = \hat{S}(-2/3;\delta) - E(T)^2, \tag{2.6.25b}$$

or

$$E(T) = (\frac{3}{2} - \ln 2) - \frac{3}{16}\delta^2 + O(\delta^4), \tag{2.6.26a}$$

$$\mathrm{Var}(T) = (-\frac{9}{16} + \frac{13}{8}\ln 2)\delta^2 + O(\delta^4). \tag{2.6.26b}$$

To verify this result a numerical simulation of (2.6.15) was carried out. We analyzed the stochastic difference equations

$$X(t+h) = X(t) + \frac{h}{\epsilon}\{Y(t) - \frac{1}{3}X^3(t) + X(t)\} \tag{2.6.27a}$$

$$Y(t+h) = Y(t) - hX(t) + \delta\sqrt{h}\,G(t), \tag{2.6.27b}$$

where $G(t)$ is a generator of random numbers with a normal distribution $\mathcal{N}(0,1)$. For $\epsilon=0.1$ and $\delta=0.25$ we computed mean and variace of 115 interjump times. The deterministic interjump time takes the value

$$T_0(\epsilon) = 1.435 \quad \text{for } \epsilon=0.1. \tag{2.6.28}$$

The expected value of T and its variance

$$E(T) = T_0(\epsilon) - 0.012, \quad \mathrm{Var}(T) = 0.034 \tag{2.6.29}$$

agree with the mean and variance of the simulation values within 0.001.

The above computations suggest that the probability density function of the interjump time can be approximated by a normal distribution. This is indeed the case for δ sufficiently small. However, for larger values of δ the normal distribution will allow negative values for the interjump time, which is physically impossible. In the statistical theory of lifetimes the inverse Gaussian distribution is used in a comparable situation, see e.g. Martz and Waller (1982). Thus the interjump times are assumed to satisfy the probability density function

$$f(t;\mu,\lambda) = (\frac{\lambda}{2\pi t^3})^{1/2} \exp\{\frac{-\lambda(t-\mu)^2}{2\mu^2 t}\} \tag{2.6.30}$$

with expected value and variance

$$E(T) = \mu, \quad \mathrm{Var}(T) = \mu^3/\lambda. \tag{2.6.31}$$

If in (2.6.17a) X_- were a positive constant, this distribution function would be exact, see Karlin and Taylor (1975).

In fig.2.6.5 the distribution of 197 interjump times is given for (2.6.27) with

$$\epsilon = 0.1 \text{ and } \delta = 0.75.$$

For the corresponding system of stochastic differential equations the asymptotic theory yields

$$E(T) = T_0(\epsilon) - \frac{3}{16}\delta^2 = 1.330$$

and

$$\mathrm{Var}(T) = (-\frac{1}{16} + \frac{13}{8}\ln 2)\delta^2 = 0.317.$$

Consequently, in the inverse Gaussian distribution function (2.6.30) we have to take

$$\mu = 1.330 \quad \text{and} \quad \lambda = 7.422.$$

In fig. 2.6.5 this distribution is given as a solid line.

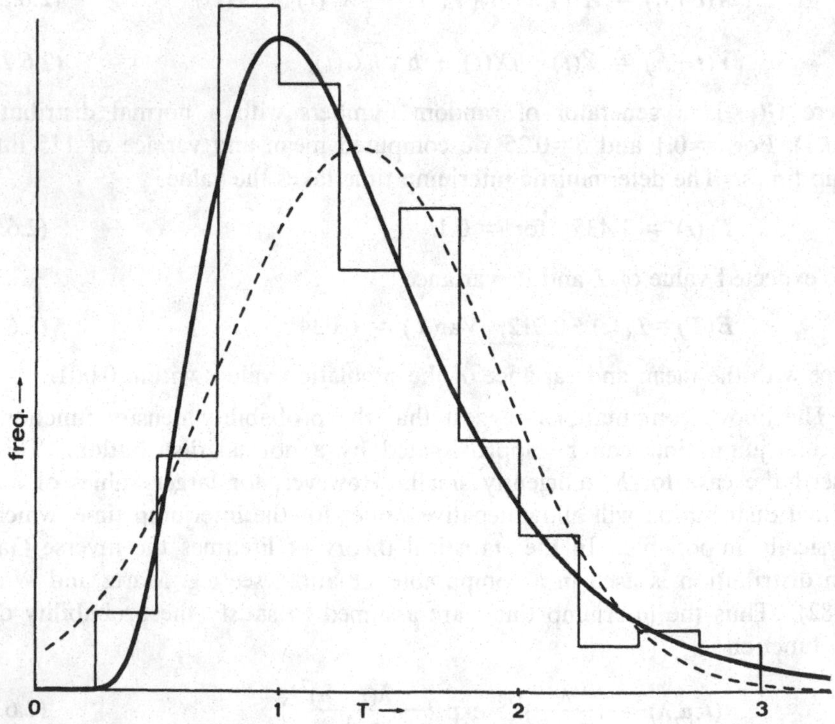

Fig.2.6.5 Distribution of interjump times from a simulation run for the stochasticly forced Van der Pol oscillator and its approximation by a normal distribution (dotted line) and by an inverse Gaussian distribution (solid line)

2.6.4 Distinction between chaos and noise

In the foregoing sections we analyzed the dynamics of explicitly given oscillators which either had a white noise input or possessed a random behaviour from themselves. From physical systems the exact equations are not always known, so that one has to decide from an output signal about the internal dynamics of such a system. In order to carry out such an analysis, values of output variables are registered at discrete times. These values can be plotted against the values at the preceding time step yielding an iteration mapping of the system. One may also study the power spectrum of the output signal to decide about the internal dynamics. Recently a method has been developed to compute Lyapunov exponents from the output signal, see Wolf *et al.* (1985).

(a) $\delta=0$

(b) $\delta=0.3$

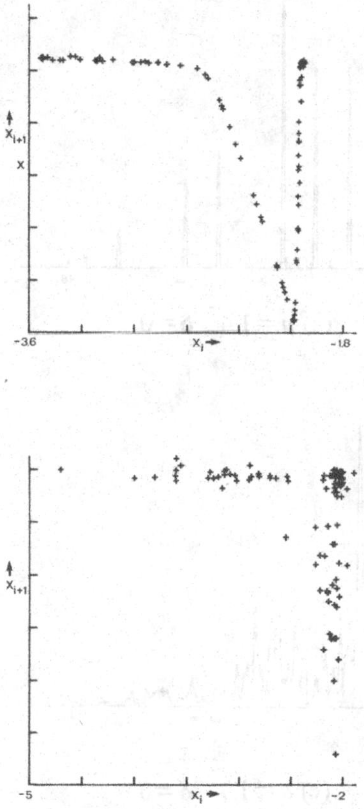

Fig.2.6.6 Iteration mapping of (2.6.32) with $a=1.7$. The quantity X_i is the lowest value of X in a cycle and X_{i+1} is the lowest value in the next cycle.

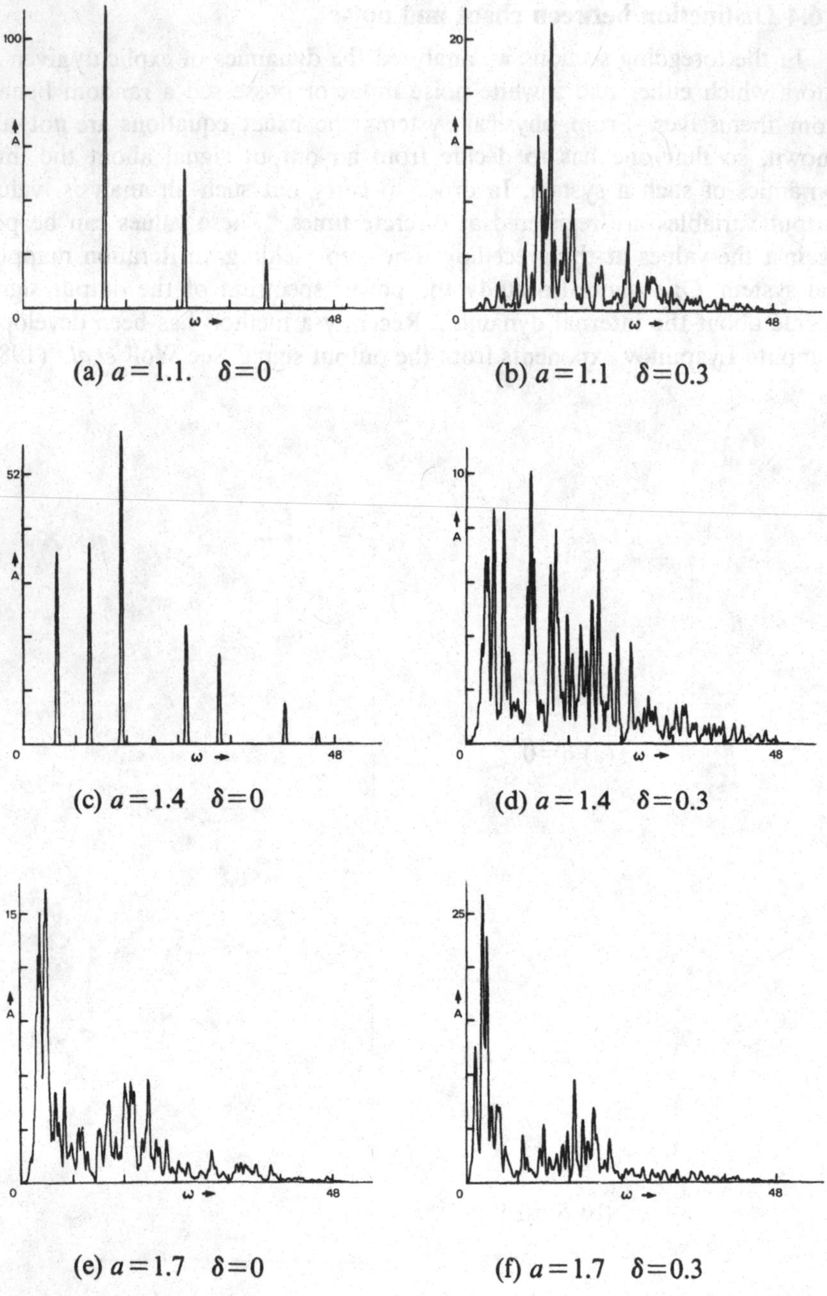

(a) $a=1.1$. $\delta=0$ (b) $a=1.1$ $\delta=0.3$

(c) $a=1.4$ $\delta=0$ (d) $a=1.4$ $\delta=0.3$

(e) $a=1.7$ $\delta=0$ (f) $a=1.7$ $\delta=0.3$

Fig.2.6.7 Power spectra of (2.6.32)

To get an idea of the qualitative differences between the effect of external noise and internal chaos we consider the system of stochastic differential equations

$$\epsilon dX=(Y-\frac{1}{3}X^3+X)dt, \quad \epsilon=0.05, \tag{2.6.32a}$$

$$dY=(-X-X^2Z)dt+\delta dW, \tag{2.6.32b}$$

$$dZ=(X+a)Z^2 dt, \tag{2.6.32c}$$

where $W(t)$ is a Wiener process. The system (2.6.32) is the chaotic Van der Pol oscillator (2.6.3) stochasticly perturbed in its second component. From the first component we take a time series of 5000 points with a time step of 0.01. In fig. 2.6.7 the power spectrum of this series for different values of a and for $\delta=0$ (no noise) and $\delta=0.3$ is given. It is noticed that periodic signals ($a=1,1$ and 1,4) without noise are easily recognized. A periodic signal with noise ($a=$ 1,4 and $\delta=0.3$) is hardly distinguished from a chaotic signal ($a=1.7$). Another method of analyzing the signal is presented in fig. 2.6.6. It gives the iteration

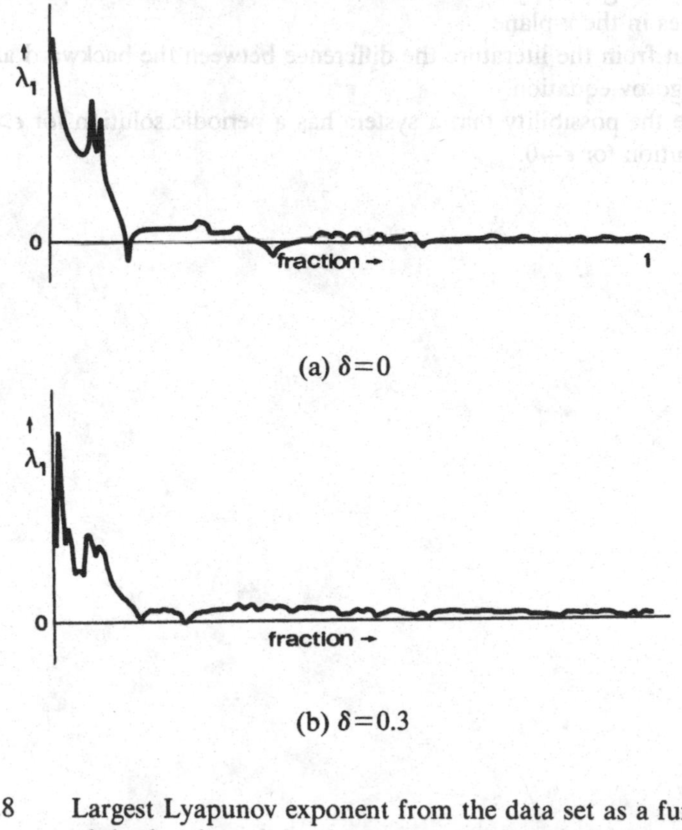

(a) $\delta=0$

(b) $\delta=0.3$

Fig.2.6.8 Largest Lyapunov exponent from the data set as a function of the fraction of the time series that is used in the approximation process mapping for the lowest value of X over each cycle. For $a = 1.7$ the points form a Cantor set, see fig. 2.6.6a. For small noise a cloud of points around this set is expected. In fig. 2.6.6b we observe a break down of this cloud for $\delta = 0.3$. The system is more or less periodic with some stochastic distribution. A second look at the power spectra confirms this observation: the peaks are higher and the area below the curve has relatively decreased. The result agrees with the outcome of the paper of Matsumoto (1984), inwhich it is stated that noise may regularize chaotic dynamics. Finally, in fig. 2.6.8 we plot the largest Lyapunov exponent as a function of the proportion of the time series that is used in the approximation process, see Wolf *et al.* (1985). It is noted that the process converges to a value close to the value 0.09 computed in section 2.6.1 and that noise hardly influences the value of the largest Lyapunov exponent.

Exercises

2.6.1 Use a computer to experiment with Eq. (2.6.2) for different values of a and δ.
2.6.2 Analyse the geometry of the manifold $\mathcal{F}$ for (2.6.3) and the projection of the trajectories in the y-plane.
2.6.3 Find out from the literature the difference between the backward and forward Kolmogorov equation.
2.6.4 Analyse the possibility that a system has a periodic solution for $\epsilon > 0$ and a chaotic solution for $\epsilon \to 0$.

3. FORCED OSCILLATION AND MUTUAL ENTRAINMENT

In this chapter we consider Van der Pol type relaxation oscillators. The coupling of these oscillators is through the second equation and may be with or without delay. A rigorous theory for the existence of entrained solutions of systems without delays is given. Numerical solutions of the iteration mapping for the phase functions of the oscillators yields results that can be used for the interpretation of entrainment phenomena in biological systems.

3.1 MODELING COUPLED OSCILLATIONS

In section 3.1.1 we discuss the modeling of periodic processes by relaxation oscillators. Next, in section 3.1.2, it is shown how the rigorous analysis of section 3.2 is directed in such a way, that the method of Mishchenko and Rosov (see section 2.1.4) can be taken as a theoretical basis. Furthermore, the formal asymptotic approximation of an entrained solution is sketched.

3.1.1 Oscillations in the applied sciences

In sections 1.6 and 1.7 we already mentioned some applications of coupled relaxation oscillators. In this chapter we give special attention to applications in the theory of modeling biological oscillations. Coupled oscillators also play an important role in the technical sciences. However, in most cases these oscillations are almost harmonic, which is not within the scope of this book. The theory of almost linear oscillations is more or less settled, see Bogoliubov and Mitropolsky (1961), Minorsky (1962) and Nayfeh and Mook (1979). It is technical applications are found in electronic networks, (Hayashi, 1964), power system networks (Anderson and Fouad, 1977), flow-induced vibrations (Blevins, 1977), wave phenomena in fluid and plasma (Whitham, 1974 and Brandstatter, 1963) and in mechanical systems (Hagendorn, 1981).

Although almost harmonic oscillations and relaxation oscillations are quantitatively not comparable, we observe some points of agreement in the qualitative behaviour of the two types of oscillation, see also Wever (1965). In both cases oscillators may get entrained: mutual entrainment in an electric power network has properties that are similar to that of a system of cardiac cells, where the first can be classified as a system of almost linear oscillators and the latter as one with highly nonlinear oscillators. Another example of agreement between the two is found in the modeling of the gastro-intestinal tract. Since the registered electrical currents are almost sinusoidal, Linkens (1979) modeled this biological system by coupled almost harmonic oscillators. However, it is not improbable that the complete electro-chemical physiological process is highly nonlinear. Modeling based upon this assumption, see section 3.4.2, yields in essence the same qualitative results.

In favour of the standpoint that biological oscillators are highly nonlinear we use the following argument. For the proper functioning of a biological system, going through a cyclic process, it is necessary to have exact information about its actual phase. This can only be achieved when the oscillator has a high orbital stability. Moreover, this cyclic process must easily be slowed down or speeded up by external periodic inputs in order to remain in phase with the environment. Exactly, these properties are found in relaxation oscillators, as we discussed in section 1.6. In general, for a biological system not all physical variables are known. In the mathematical analysis one uses a prototype of oscillator with the two properties specified above. This can be a radial isochron clock, see Hoppensteadt and Keener (1982), or a two dimensional dynamical system, such as the Van der Pol relaxation oscillator. In this chapter we investigate the consequences of choosing the latter one. The model of Van der Pol and Van der Mark (1928) of the heart-beat can be seen as the starting point of this method of analyzing biological oscillations, see fig. 3.1.1.

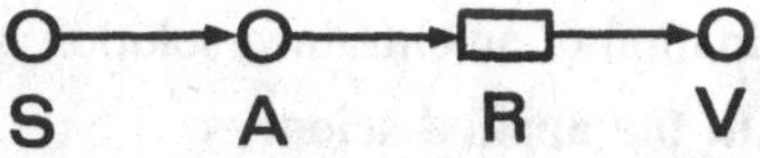

Fig.3.1.1 Schematic representation of the heart by three coupled relaxation oscillators: S(=Sinus), A(=Atria) and V(=Ventricles). R is a delay system representing the finite time interval needed to pass the bundle of His. The coupling between A and V can be varied giving arise to irregular rhythms of the system comparable with cardiac arrythmias.

3.1.2 The system of differential equations and the method of analysis

In Eq (1.4.1) we gave the general form of a singularly perturbed system of differential equations that may have a periodic solution being a relaxation oscillation:

$$\epsilon\frac{dx_i}{dt} = F_i(x,y), \quad i = 1,...,p, \tag{3.1.1a}$$

$$\frac{dy_j}{dt} = G_j(x,y) \quad j = 1,...,q, \tag{3.1.1b}$$

where $0<\epsilon<<1$. When such a system represents n coupled 2-dimensional oscillators, it has, in case of coupling by the slow component, the form

$$\epsilon\frac{dx_i}{dt} = F_i(x_i,y_i), \tag{3.1.2a}$$

$$\frac{dy_i}{dt} = G_i(x_i,y_i;\delta)+\delta\sum_{j\neq i}H_{ij}(x_i,y_i,x_j,y_j), \quad i = 1,...,n, \tag{3.1.2b}$$

where $0<\delta<<1$. The reason for choosing weak coupling by the slow component lies in the fact that this will not perturb the limit cycle of the individual oscillators, but only the phase velocity, which may get entrained by the coupling. This meets the requirement that biological oscillators remain at the limit cycle and easily adapt their phase velocity. In our asymptotic analysis of the problem it is assumed that

$$0<\epsilon<<\delta<<1. \tag{3.1.3}$$

Mishchenko and Rosov (1980) analyse relaxation oscillations of (3.1.1) with and $0<\epsilon<<1$, see section 2.1.4. For our type of system (3.1.2) we have to prove the convergence of the regular asymptotic expansion with respect to δ in order to apply their method of constructing asymptotic approximations of periodic solutions for ϵ small. This part of our analysis is carried out in sections 3.2.1-3.2.2.

In the remaining part of this section we give the formal construction of an asymptotic approximation. The relaxation oscillators we consider are, more specifically, of the type

$$\epsilon\frac{dx}{dt} = y - F(x), \tag{3.1.4a}$$

$$\frac{dy}{dt} = a - x. \tag{3.1.4b}$$

The discontinuous approximate solution ($\epsilon=0$) satisfies

$$F'(X_0)\frac{dX_0}{dt} = a - X_0. \tag{3.1.5}$$

We take into consideration the periodic forcing of the relaxation oscillator (3.1.4) through its y-component:

$$\epsilon dx/dt = y - F(x), \tag{3.1.6a}$$

$$dy/dt = a - x + \delta h(t), \quad h(t+T) = h(t) \tag{3.1.6b}$$

with $h(t)$ a piecewise continuous function. For $\epsilon\to 0$ the trajectories satisfy $y=F(x)$ or $y=$constant so that the forcing term h will not change the closed trajectory in the phase plane. It may only influence the velocity of the oscillator on the limit cycle. Consequently, a solution of (3.1.6) is approximated by

$$x = X_0(\phi(t)), \quad y = Y_0(\phi(t)), \tag{3.1.7}$$

where $(X_0(t), Y_0(t))$ represents a discontinuous approximation of the free oscillator; see (3.1.5). Substitution in (3.1.6) for $\epsilon=0$ yields

$$\frac{dY_0}{d\phi}\frac{d\phi}{dt} = a - X_0(\phi(t)) + \delta h(t) \tag{3.1.8}$$

or, using (3.1.4b),

$$\frac{d\phi}{dt} = 1 + \frac{\delta h(t)}{a - X_0(\phi(t))}, \quad \phi(0) = \alpha^{(0)}. \tag{3.1.9}$$

Integration gives the following approximation valid for bounded t:

$$\phi(t) = \alpha^{(0)} + t + \delta \int_0^t \frac{h(\bar{t})}{a - X_0(\alpha^{(0)} + \bar{t})} d\bar{t} + O(\delta^2). \tag{3.1.10}$$

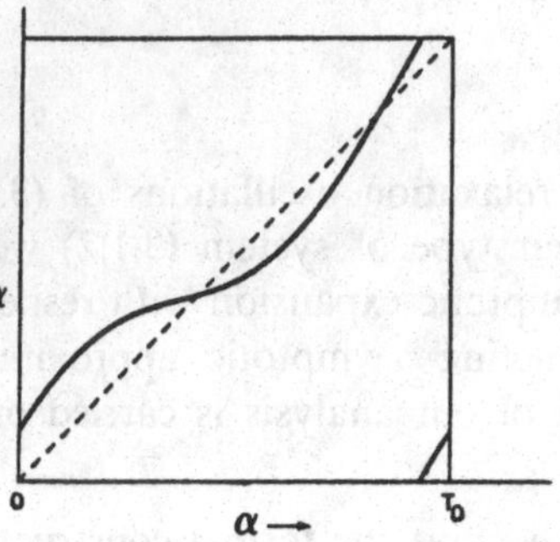

(a) The iteration map P

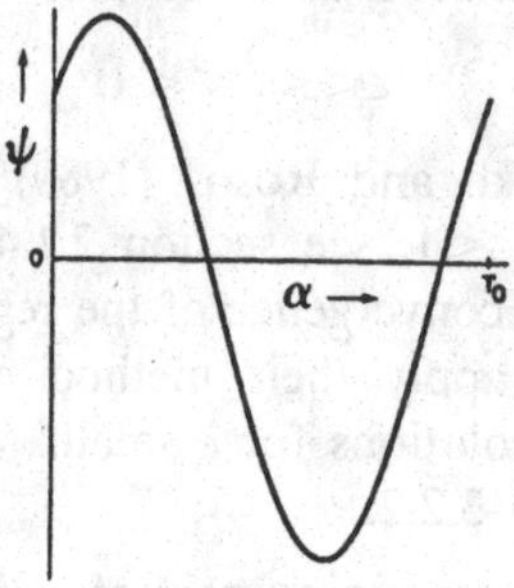

(b) the phase shift function ψ

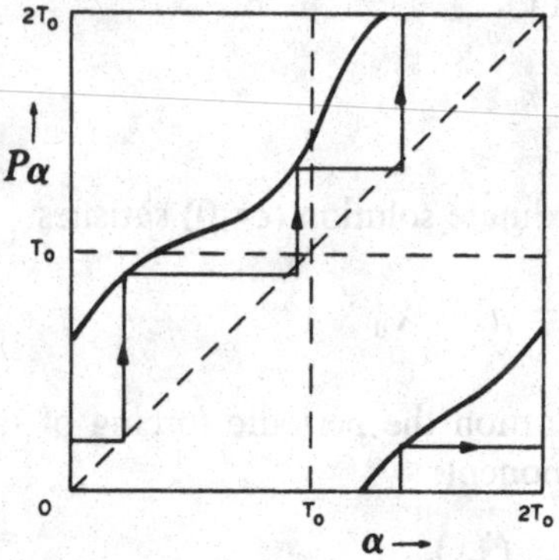

(c) a higher order fixed point of the map P

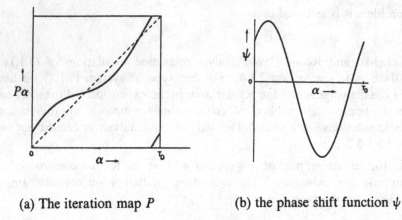

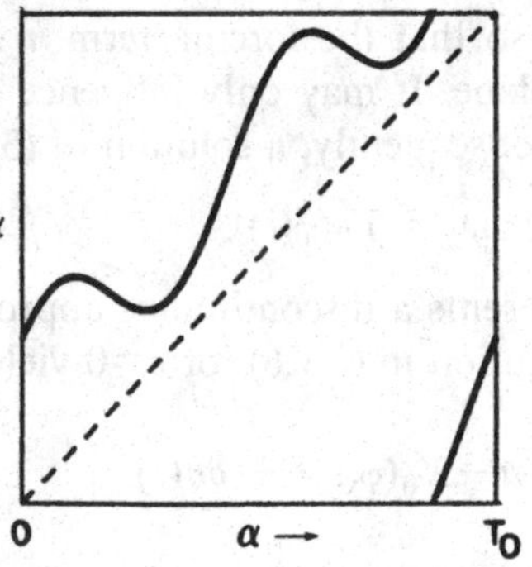

(d) an iteration map with chaotic dynamics

Fig.3.1.2 Properties of the iteration mapping

Over one period T the forcing causes a phase shift $\delta\psi(\alpha^{(0)})$ with

$$\psi(\alpha) = \int_0^T \frac{h(\bar{t})}{a - X_0(\alpha+\bar{t})} d\bar{t}. \tag{3.1.11}$$

Considering the value of ϕ at times $t=kT$, we obtain the iteration map P for the phase $\alpha^{(k+1)}$ or in a explicit form with accuracy $O(\delta^2)$:

$$\alpha^{(k)} = \alpha^{(k)} + T + \delta\psi(\alpha^{(k)})(\text{ mod})T_0, \tag{3.1.12}$$

where T_0 is the period of the autonomous system in the limit $\epsilon \to 0$. From the iteration map we analyse the limit behaviour of the system. In the simplest case it has a stable fixed point that corresponds with a periodic solution of period T. Other possibilities are higher stable subharmonic solutions (see e.g. fig. 3.1.2c) and chaotic solutions for $\delta = O(1)$, see fig. 3.1.2.d and Guckenheimer (1980). Clearly a fixed point $\bar{\alpha}$ satisfies

$$\psi(\bar{\alpha}) = (mT_0 - T)/\delta$$

for some integer m and is stable if $\psi'(\bar{\alpha}) < 0$. Phase locking will occur if

$$\min_{\alpha} \delta\psi(\alpha) < mT_0 - T < \max_{\alpha} \delta\psi(\alpha).$$

Coupled Oscillators. We are now in the position to handle systems of coupled relaxation oscillators satisfying

$$\epsilon \frac{dx_i}{dt} = y_i - F_i(x_i) \tag{3.13a}$$

$$\frac{dy_i}{dt} = c_i(\delta)(a_i - x_i) + \delta \sum_{j \neq i} h_{ij}(x_{i'} y_{i'} x_{j'} y_{j'}), \tag{3.1.13b}$$

where h_{ij} is assumed to be continuous. Each oscillator describes a free oscillation given by $(X_{i0}(\phi_{i0}(t)), Y_{i0}(\phi_{i0}(t))$, with

$$\phi_{i0}(t) = \alpha_i^{(0)} + c_i(\delta)t. \tag{3.1.14}$$

In case the oscillators are coupled, the phase functions are approximated by

$$\phi_i(t) = \phi_{i0}(t) + \delta \sum_{j \neq i0} \int^t \frac{h_{ij}(\phi_{i0}(\bar{t}), \phi_{j0}(\bar{t}))}{a_i - X_i(\phi_{i0}(\bar{t}))} d\bar{t} + O(\delta^2). \tag{3.1.15}$$

Let us assume that the unperturbed oscillators $(\delta=0)$ have autonomous periods $T_{i\epsilon}$ satisfying

$$T_{10} : T_{20} : \ldots : T_{n0} = j_1 : j_2 : \ldots : j_n, \tag{3.1.16}$$

where $j_i, i=1,\ldots,n$ are integers. The fact that $c_i(\delta) = c_i(0) + O(\delta)$ results in autonomous periods of the perturbed system that may differ $O(\delta)$ from this ratio. Next we introduce the common unperturbed period T, the smallest number for which the quotients $T/T_{i0}, i=1,\ldots,n$ are positive integers.
The phase shift function is defined by

$$\psi_i(\alpha) = \sum_{j\neq i} \psi_{ij}(\alpha_i,\alpha_j), \tag{3.1.17}$$

$$\psi_{ij}(\alpha_i,\alpha_j) = \int_0^T \frac{h_{ij}(\alpha_i + c_i(\delta)t, \alpha_j + c_j(\delta)t)}{X_i(\alpha_i + c_i(\delta)t)} dt, \quad i\neq j.$$

For the iteration map P of the phases $\alpha_i^{(k)}$ at times $t=kT$ we obtain

$$\alpha_i^{(k+1)} = \alpha_i^{(k)} + c_i(\delta)T + \delta\psi_i(\alpha_i^{(k)}) \,(\mathrm{mod}) T_0 \quad \text{for} \quad i=1,\ldots,n$$

or

$$\alpha^{(k+1)} = P\alpha^{(k)}.$$

More specifically, the phase shift function ψ_{ij} depends upon $\beta_{ij}=\alpha_i-\alpha_j$, as seen from (3.1.17) by shifting the integration interval over α_j. If we set $\alpha_1=0$, then all phase differences β_{ij} are uniquely determined from the remaining $n-1$ phases α_j. The system (3.1.13) has a periodic solution with a period of about T if the following system of n algebraic equations for $\alpha_2,\ldots,\alpha_n$ and q has a solution:

$$c_i(\delta)T + \delta\,\psi_i(\alpha) = \delta q (\mathrm{mod}) T_0, \; i = 1,\ldots,n. \tag{3.1.18}$$

The period of the approximation for $\epsilon\to 0$ takes the value $T+\delta q$.

In section 3.2.4 this formal construction is extended to oscillators coupled with delay, while in the sections 3.3 and 3.4 various applications are worked out.

Exercises 3.1.1 Analyse the entrainment of an almost linear oscillator forced by an harmonic oscillator, see the literature mentioned in the text.

3.1.2 For the approximation (3.1.7) the parameter δ does not have to be small. However, for large δ with $\delta\epsilon\to\infty$ a trajectory may leave the limit cycle. Where does this occur?

3.1.3 Analyse the possibility of chaotic solutions of (3.1.12) for $\delta=O(1)$, see Guckenheimer (1980).

3.1.4 Investigate the relation between the phase shift function and the method of averaging by introduction of $\Phi=\phi-t$ in (3.1.9), see e.g. Nayfeh and Mook (1979).

3.2 A RIGOROUS THEORY FOR WEAKLY COUPLED OSCILLATORS

In this section we investigate a finite system of coupled relaxation oscillators, described by the following differential equations:

$$\epsilon\frac{dx_i}{dt} = y_i - F_i(x_i), \tag{3.2.1a}$$

$$\frac{dy_i}{dt} = c_i(\delta)(a_i - x_i) + \delta h_i(x,y),\ i = 1,...,n, \tag{3.2.1b}$$

with n the number of oscillators and $x = (x_1,...,x_n)$, $y = (y_1,...,y_n)$. The function h_i represents the coupling between the oscillators. It is assumed that F_i and h_i are C^∞, unless it is specified otherwise, and that, moreover, a set of n integer numbers j_i exists such that

$$T_{10}/c_1(0): ... :T_{n0}/c_n(0) = j_1: ... :j_n. \tag{3.2.1c}$$

It is noted that this system has the form of a coupled *CD*-system, see (1.6.3).

The validity of asymptotic approximations of entrained solutions of (3.2.1) is proved in sections 3.2.1-3.2.3 (Jansen, 1978). In section 3.2.4 a formal extension to oscillators coupled with delay is made.

3.2.1 Validity of the discontinuous approximation

We shall investigate the behaviour of the solution of (3.2.1) for ϵ tending to zero. Just as in the previous chapter we introduce the reduced system.

$$y_i = F_i(x_i), \tag{3.2.2a}$$

$$\frac{dy_i}{dt} = c_i(\delta)(a_i - x_i) + \delta h_i(x,y), \quad i = 1,...,n. \tag{3.2.2b}$$

Combination of these equations yields a reduced system

$$F_i'(x_i)\frac{dx_i}{dt} = c_i(\delta)(a_i - x_i) + \delta h_i(x,F(x)), \ \ i=1,...,n. \tag{3.2.2c}$$

We also introduce the reduced fast system

$$\epsilon\frac{dx_i}{dt} = y_i - F_i(x_i), \ \ i=1,...,n \tag{3.2.3}$$

in which the constants y_i can be considered as parameters.

The limit solution of (3.2.1) is defined as follows. When (x,y) is not a stable equilibrium point of the fast system an instantaneous jump is made along a trajectory of the fast equation until a stable equilibrium of this equation is reached (a return point). Afterwards the singular solution satisfies the reduced system until one or more of the variables y_i reaches a local extremum of F_i. At that point (a leaving point) the reduced equation cannot be satisfied any more. Then the limit solution makes an instantaneous jump along the unique trajectory of the fast equation departing from the leaving point, until a new return point is reached. After the jump the limit solution is described again by the reduced system. The limit solution approximates the exact solution in the sense that the trajectories of the solution of (3.2.1) will tend to the trajectories of the limit solution of (3.2.1) if ϵ tends to zero. Moreover, system (3.2.1) has a periodic solution if it has a periodic limit solution satisfying certain conditions, see section 2.1.4. We have to introduce some concepts before formulating a theorem establishing this behaviour.

Let the regular parts of the closed trajectory $(x_{i0}(t), y_{i0}(t))$ of the periodic singular solution of (3.1.1) be indicated by L_i. Then we may define the following n-dimensional surface in the space $\mathbb{R}^{2n}$.

$$L^n = \{(x,y) \in \mathbb{R}^{2n} \mid (x_i,y_i) \in L_i \text{ for } i = 1,...,n\}. \tag{3.2.4}$$

This means that $(x,y) \in L^n$ when each oscillator (x_i,y_i) lies on the regular trajectory L_i of the singular solution of one isolated oscillator. We shall show that for δ sufficiently small a singular solution $(x(t),y(t))$ will remain in the set L^n once it has arrived there.

Lemma 3.2.1. When δ is sufficiently small but finite, the set L^n is invariant with respect to singular solutions of (3.2.1).

Proof. Let $(x,y) \in L^n$. We first consider the case where

$$F'_i(x_i) \neq 0, \quad i = 1,...,n.$$

It follows from (3.2.2) that for δ sufficiently small

$$y_i = F_i(x_i) \tag{3.2.5a}$$

$$\text{sign}\,(dx_i/dt) = -\text{sign}\,(x_i - a_i) \tag{3.2.5b}$$

which implies that the coupled oscillators run through the regular parts in the same direction as the uncoupled oscillator. In the leaving points, where the derivative of $F_i(x_i)$ is zero for some i, the system makes an instantaneous jump to a new stable equilibrium of (3.2.3). This is the same for coupled oscillators. Consequently, L_n is invariant. □

For a concise formulation of a theorem on periodic solutions of (3.2.1) we will need the following two definitions.

Definition 3.2.1. Let $n \geqslant 2$ and let W be a smooth $(n-1)$- dimensional surface lying in the n-dimensional surface L^n. Moreover, let W be nowhere tangent to the trajectories of the limit solutions of (3.2.1) (transversal intersection). Let w be a point of W and let a singular solution start in w. Then it may happen that this limit solution will return in W. If so, denote the point of first return by $\mathcal{P}(w)$. In this way a mapping, $\mathcal{P}$, is defined from a part of W into W. This mapping is called the Poincaré map of W produced by the limit solution.

It is clear that there exists a periodic limit solution if $\mathcal{P}$ has a fixed point, i.e. a point $w \in W$ such that $\mathcal{P}(w) = w$. The closed trajectory of such a periodic solution will be indicated by Z_0, the period by $T(0)$.

Definition 3.2.2. A periodic limit solution of (3.2.1) will be called C-stable if a surface W exists as described in definition 3.2.1, such that the corresponding Poincaré map $\mathcal{P}$, is contracting at the intersection of W and Z_0.

The following theorem by Mishchenko and Rosov (1980), see section 2.1.4, permits us to fix our attention to C-stable periodic singular solutions of Eq. (3.2.1).

Theorem 3.2.1. Let δ be such that Eq. (3.2.1) has a C-stable periodic singular solution with trajectory Z_0 and period $T(0)$. Let only one of its components $x_i(t), i = 1,2,...,n$ be discontinuous at a time. Then a positive function $\bar{\epsilon}(\delta)$ exists such that, for $0<\epsilon\leqslant\bar{\epsilon}(\delta)$, Eq. (3.2.1) has a periodic solution with period $T(\epsilon)$ and trajectory $Z_\epsilon \subset \mathbb{R}^{2n}$ satisfying:

$$(i) \;\; Z_\epsilon \to Z_0 \;\; \text{for} \;\; \epsilon \to 0$$

and

$$(ii) \;\; T(\epsilon) = T(0) + O(\epsilon^{2/3}).$$

3.2.2 Construction of the asymptotic solution

We now introduce the phase-map

$$\Phi : \mathbb{R}^n \to L^n \tag{3.2.6a}$$

defined by

$$\Phi : (\phi_1,...,\phi_n) \to (X_{10}(\phi_1),...,X_{n0}(\phi_n), Y_{10}(\phi_1),..., Y_{n0}(\phi_n)). \tag{3.2.6b}$$

where $(X_{i0}(t)), Y_{i0}(t))$ is the limit solution of the decoupled oscillator for $\epsilon,\delta \to 0$. The point $\phi_i \in \mathbb{R}$ will be called the *phase* of oscillator i. It is easy to see that each point of L^n has an original in $\mathbb{R}^n$, that is: Φ is surjective. The map Φ is T_{i0}-periodic in its arguments and locally invertible. Moreover, it follows from (3.2.2c) that it is locally diffeomorphic except at the jump hyperplanes.

Lemma 3.2.2. A limit solution of Eq. (3.2.1), with initial value in L^n, is represented by

$$x_i(t) = X_{i0}(\phi_i(t)), \; y_i(t) = Y_{i0}(\phi_i(t)), \;\; i = 1,...,n, \tag{3.2.7a}$$

where the functions $\phi_i(t)$ satisify the phase equation

$$\frac{d\phi_i}{dt} = c_i(\delta) + \delta \frac{h_i[\phi(t)]}{a_i - X_{i0}(\phi_i(t))}. \tag{3.2.7b}$$

Proof. Representation (3.2.7a) follows from the invariance of L^n and from the

surjectivity of Φ. Eq. (3.2.7b) is obtained by substituting (3.2.7a) into (3.2.2c).
□

We now approximate the solution of (3.2.7b) in order to investigate the mapping $\mathcal{P}$. Since δ is a small parameter it is natural to solve (3.2.7b) by iteration. Let $\phi(0) = \alpha$. Then the first and second iterates are

$$\phi_{i0}(t) = \alpha_i + c_i(0)t, \tag{3.2.8a}$$

$$\hat{\phi}_i(t) = \alpha_i + c_i(\delta)t + \delta \int_0^t \frac{h_i[\phi_0(\bar{t})]}{a_i - X_{i0}(\phi_{i0}(\bar{t}))} d\bar{t}. \tag{3.2.8b}$$

The integral on the right-hand side of (3.2.8b) exists since the denominator is bounded away from zero.

We frequently impose a specific condition upon the initial value $\phi(0) = \alpha$. Therefore we state:

Definition 3.2.3. A point $\alpha \in \mathbb{R}^n$ will be called *regular* if the functions $X_{i0}(\alpha_i + c_i(0)t)$, $i = 1,2,\dots,n$ are continuous in $t = 0$ and if they become discontinuous one at a time.

Lemma 3.2.3. Let α be regular. Then Eq. (3.2.7b) with $\phi(0) = \alpha$ has a unique solution $\phi(t)$. Moreover, we have that

$$\hat{\phi}_i(t) = \phi_i(t) + O(\delta^{3/2}) \quad (t \text{ bounded}). \tag{3.2.9}$$

Proof. Lemma 3.2.2 shows that there is a one to one correspondence between solutions of (3.2.1) lying in L^n and solutions of (3.2.7b). The singular solution of (3.2.1) with initial value $\phi(\alpha) \in L^n$ exists and is unique by definition. This solution remains in L^n by lemma 3.2.1. It follows that (3.2.7b) has a unique solution.

The correctness of (3.2.9) remains to be proved. We start with an investigation of the regularity of the function $h[\phi]$. It can be shown that $X_{i0}(\phi_{i0}(t))$ is piecewise Hölder continuous with exponent 1/2 and has finite jumps. This property of $h[\phi]$ may be used to estimate $\phi(t) - \hat{\phi}(t)$ which according to (3.2.7b) and (3.2.8), can be written as

$$\phi(t) - \hat{\phi}(t) = \delta \int_0^t f(\phi(\bar{t})) - f(\phi_0(\bar{t})) d\bar{t}. \tag{3.2.10}$$

It follows from (3.2.7) and (3.2.8) that

$$|\phi(t) - \phi_0(t)| = O(\delta)(t \text{ bounded}).$$

When $\phi(t)$ and $\phi_0(t)$ are at the same side of a jump point

$$|f(\phi(t)) - f(\phi_0(t))| \leqslant k|\phi(t) - \phi_0(t)|^{\frac{1}{2}} = O(\delta^{\frac{1}{2}}), \tag{3.2.11}$$

in which k is the Hölder constant of f. This gives an $O(\delta^{3/2})$ contribution to (3.2.10). However $\phi(t)$ and $\phi_0(t)$ may be located at different sides of a jump point in which case

$$|f(\phi) - f(\phi_0)| = O(1).$$

Using

$$|\phi(t) - \phi_0(t)| = O(\delta)$$

we conclude that $\phi(t)$ and $\phi_0(t)$ remain at different sides for $O(\delta)$ time intervals. These give an $O(\delta^2)$ contribution to (3.2.10). The correctness of (3.2.9) follows immediately. □

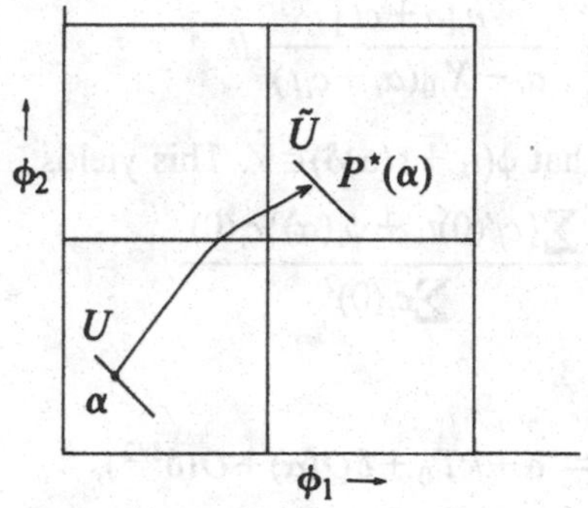

Fig.3.2.1 The path in phase space for a system of two coupled oscillators. In first order approximation the trajectories satisfy $d\phi/dt = c(0)$.

Using approximation (3.2.10) we analyse the properties of the Poincaré map $\mathcal{P}$ of definition 3.2.1. Let V be an $(n-1)$-dimensional plane in $\mathbb{R}^n$ perpendicular to the vector $c(0) = (c_1(0),...,c_n(0))$, see fig.3.2.1. Let α be a regular point in V and let U be a neighbourhood of α in V. We denote by $\tilde{V}$ and $\tilde{U}$ translations of V and U of kT_0, with k such that $\alpha + kT_0$ is contained in $\tilde{V}$. Such an integer number k always exists because of condition (3.2.1c). Furthermore, we define $W = \Phi(U) = \Phi(\tilde{U})$, see (3.2.6b). It follows from lemma 3.2.2 that the Poincaré map that corresponds with $W \subset L^n$ is given by

$$\mathcal{P} = \Phi \mathcal{P}^* \Phi^{-1}, \tag{3.2.12}$$

where $\mathcal{P}^*$ is the mapping from U into $\tilde{U}$ as given by (3.2.7b).

Lemma 3.2.4. The mapping $\mathcal{P}$ has a fixed point $\Phi(\overline{\alpha})$ if there exists a regular point $\overline{\alpha}$ such that

$$\mathcal{P}^*(\overline{\alpha}) = \overline{\alpha} + kT_0.$$

Moreover $\mathcal{P}$ is contracting in $\Phi(\overline{\alpha})$ if the eigenvalues of the derivative of $\mathcal{P}^*$ in $\overline{\alpha}$ have absolute values less than one.

Proof. The first part of the lemma follows immediately from representation (3.2.12) and from the fact that Φ is T_0 periodic. The second part is proved by defining a distance in $W \subset L^n$ as the corresponding Euclidean distance in $U \subset \mathbb{R}^n$, produced by the mapping Φ^{-1}. □

From (3.2.8b) it follows that for $t = t_k + s\delta$ with

$$t_k = kT_{i0}/c_i(0) = T(0)$$
$$\phi_i(t_k + s\delta) = \alpha_i + c_i(\delta)(t_k + s\delta) + \delta\psi_i(\alpha) + O(\delta^{3/2}),$$

where

$$\psi_i(\alpha) = \int_0^{T(0)} \frac{h_i[\alpha + ct]}{a_i - X_{i0}(\alpha_i + c_i t)} dt. \tag{3.2.13}$$

We choose $s(\alpha)$ such that $\phi(t_k + s(\alpha)\delta) \in \tilde{V}$. This yields

$$s(\alpha) = -\frac{\sum\{c_i'(0)t_k + \psi_i(\alpha)\}c_i(0)}{\sum c_i(0)^2} \tag{3.2.14}$$

and so

$$\phi(t_k + s\delta) = \alpha + kT_0 + \delta Q(\alpha) + O(\delta^{3/2}), \tag{3.2.15a}$$

$$Q(\alpha) = \psi(\alpha) + t_k\, c'(0) + s(\alpha)\, c(0). \tag{3.2.15b}$$

It is indeed so that $Q(\alpha)$ is perpendicular to $c(0)$, and thus

$$\alpha + kT_0 + \delta Q(\alpha) \in \tilde{V}. \tag{3.2.16}$$

We summarize the above calculations in the following lemma.

Lemma 3.2.5. The mapping $\mathcal{P}^*$ is approximated with an accuracy $O(\delta^{3/2})$ by the restriction to the plane U of the mapping

$$\hat{\mathcal{P}}^*\alpha = \alpha + kT_0 + \delta Q(\alpha). \tag{3.2.17}$$

3.2.3 Existence of a periodic solution

Using lemma 3.2.5 we prove next:

Lemma 3.2.6. Suppose that

(i) $\tilde{\alpha}$ is regular (regularity condition),

(ii) all eigenvalues, except one, of the derivative of $Q(\alpha)$ in $\tilde{\alpha}$ have negative real parts (stability condition).

(iii) $Q(\tilde{\alpha}) = 0$ (entrainment condition),

Then a point $\tilde{\beta} = \tilde{\alpha} + O(\delta)$ exists such that $\mathcal{P}^*(\tilde{\beta}) = \tilde{\beta} + kT_0$ with the eigenvalues of the derivative of $\mathcal{P}^*$ in $\tilde{\beta}$ having absolute values less than one.

Proof Due to the fact that the left hand side of (3.2.7b) is not Lipschitz continuous it is not immediately clear that P^* is C^∞ in α and δ. This problem can be overcome by a good choice of independent variables. Since

$$\phi(t)-\phi_0(t) = \phi(t)-(\tilde{\alpha}+tc) = O(\delta),$$

we know that $\phi(t)$ will remain in an δ-tube N_δ, around the trajectory of $\tilde{\alpha}+tc$. By the regularity of $\tilde{\alpha}$ this trajectory will not cross points where $X_0(\phi)$ is discontinuous in more than one variable. It follows that δ can be chosen so small that N_δ does not contain such points either. This implies that we know the order in which the variables $\phi_1,\phi_2,\ldots,\phi_n$ will cause the discontinuities of $X_0(\phi(t))$; this order will be indicated by $\phi_{i_1},\phi_{i_2},\ldots$. The corresponding planes of intersection of N with the discontinuity-planes of $h(\phi)$ will be denoted by $U_1,U_2,\ldots$, see fig.3.2.1.

The map

$$P^*:U\to U$$

is the composition of the mappings

$$U\to U_1, U_1\to U_2,\ldots, U_l\to U,$$

produced by the trajectories of (3.2.7b) (the plane U_l is the last of the planes $U_1,U_2,\ldots$ that is passed before U is reached). We shall show that these mappings are C^∞ with respect to δ and the initial conditions. This implies that P^* is C^∞ with respect to δ and α.

In the part of N_δ bounded by U and U_1 we take ϕ_{i_1} as independent variable, so that Eq. (3.2.7b) gets the form:

$$\frac{d\phi_i}{d\phi_{i_1}} = \frac{c_i(\delta)+\delta f_i(\phi_i)}{c_{i_1}(\delta)+\delta f_{i_1}(\phi_{i_1})}, i\neq i_1. \tag{3.2.18}$$

For δ sufficiently small this equation has bounded derivatives of any order with respect to δ and $\phi_i, i\neq i_1$. Moreover it is continuous with respect to ϕ_{i_1}.

It follows from theorem 7.5 of Chapter 1 of Coddington and Levinson (1955) that the mapping $U\to U_1$ determined by the trajectories of (3.2.7b) is C^∞ with respect to δ and the initial condition. The other mappings are treated in the same way; for the map $U_l\to U$ we take $\sum\phi_i$ as independent variable. Consequently, P^* and its derivatives with respect to α (denoted by $dP^*, d^2P^*\cdots$) can be developed in a Taylor series around $\delta = 0$. Using lemma 3.2.5 we obtain

$$P^*(\alpha) = \alpha_u+\delta Q_u(\alpha)+O(\delta^2), \tag{3.2.19a}$$

$$dP^*(\alpha) = d\alpha_u+\delta dQ_u(\alpha)+O(\delta^2), \tag{3.2.19b}$$

$$d^2P^*(\alpha) = O(\delta). \tag{3.2.19c}$$

The derivative $d(\alpha+\delta Q(\alpha))$ has an eigenvalue 1 that corresponds to the eigenvector c. According to (3.2.16) $\alpha+kT_0+\delta Q(\alpha)$ maps U into U. This implies that $d(\alpha_u+\delta Q_u(\alpha))$ is contracting if the other eigenvalues of $d(\alpha+\delta Q(\alpha))$ are lying strictly within the unit cirle. It now follows from the conditions of the lemma that $\alpha_u+\delta Q_u(\alpha)$ is contracting in $\tilde{\alpha}$. It remains to be shown that a point $\tilde{\beta}$ exists, close to $\tilde{\alpha}$, such that

$$P^*(\tilde{\beta}) = \tilde{\beta}, \tag{3.2.20}$$

with P^* contracting in $\tilde{\beta}$. Let $\beta \in U$. Then

$$P^*(\beta)-\tilde{\alpha} = P^*(\beta)-P^*(\tilde{\alpha})+O(\delta^2). \tag{3.2.21}$$

By the fact that $d(\alpha+\delta Q(\alpha))$ is contracting in $\tilde{\alpha}$, and that $d^2P^* = O(\delta)$ one has

$$|P^*(\beta)-P^*(\tilde{\alpha})| \leqslant (1-\gamma\delta)|\beta-\tilde{\alpha}|+m|\beta-\tilde{\alpha}|^2+n\delta^2 \tag{3.2.22}$$

for some positive γ, m and n. Let $|\beta-\tilde{\alpha}| \leqslant 2\delta n/\gamma$ then for δ sufficiently small

$$|P^*(\beta)-\tilde{\alpha}| \leqslant 2\delta n/\gamma.$$

Thus a δ-ball in U around $\tilde{\alpha}$ is mapped on itself by P^*. It follows that P^* has fixed point $\tilde{\beta}$ in an δ-neighbourhood of $\tilde{\alpha}$. Using $d^2P^* = O(\delta)$ we see that d is contracting in $\tilde{\beta}$. This concludes the proof of the lemma. □

At this stage our knowledge of the behaviour of the singular solution is sufficient to return to the original equation (3.2.1). Putting together the results of this section we obtain as our final result.

Theorem 3.2.2. Suppose that $\tilde{\alpha}$ satisfies the three conditions of Lemma 3.2.6. Then a positive constant $\bar{\delta}$ and a positive function $\bar{\epsilon}(\delta)$, defined on $(0,\bar{\delta})$ exist such that for $0<\delta<\bar{\delta}$ and $0<\epsilon<\bar{\epsilon}(\delta)$ Eq. (3.2.1) has a periodic solution. This periodic solution has the following properties. Its trajectory $Z_{\epsilon,\delta}$ tends to the trajectory of

$$(X_0(\tilde{\alpha}+c(0)t),\ Y_0(\tilde{\alpha}+c(0)t)) \tag{3.2.23}$$

when ϵ and δ tend to zero. Its period $T_\delta(\epsilon)$ satisfies

$$T_\delta(\epsilon) = T(0) + \delta s(\tilde{\alpha}) + O(\delta^{3/2}) + O(\epsilon^{2/3}) \tag{3.2.24}$$

with $s(\tilde{\alpha})$ given by (3.2.14).

3.2.4 Formal extension to oscillators coupled with delay

The construction of a periodic solution can be applied to oscillators coupled with delay:

$$\epsilon\frac{dx_i}{dt} = y_i - F(x_i), \tag{3.2.25a}$$

$$\frac{dy_i}{dt} = c_i(\delta)(a_i - x_i) + \delta h_i(x,y,\hat{x},\hat{y}), \quad i = 1,\ldots,n, \tag{3.2.25b}$$

where

$$\hat{x} = (x_1(t-\rho_{i1}),...,x_n(t-\rho_{in})), \tag{3.2.26a}$$

$$\hat{y} = (y_1(t-\rho_{i1},...,y_n(t-\rho_{in})). \tag{3.2.26b}$$

For the phase shift function (3.2.13) we then obtain

$$\psi_i(\alpha) = \int_0^{T_{i0}} \frac{h_i[\alpha+ct-\rho_i t]}{a_i - X_{i0}(\alpha+c_i t)} dt. \tag{3.2.27}$$

The expressions for Q and $T_\rho(\epsilon)$ do not change, see (3.2.15b) and (3.2.24). Since theorem 3.2.1 does not apply to systems with delay, the correctness of the asymptotic approximation cannot be proved in this way.

Exercises

3.2.1 In the discontinuous approximation $a_i \neq X_{i0}(t)$ for all t, so the denominator in (3.2.8b) is never zero. In the next order approximation this is not the case. What are the consequences?

3.2.2 Compare the result (3.2.24) with (3.1.18).

3.2.3 Is it possible to prove the validity of the approximation if $\bar{\alpha}$ is not regular?

3.3 COUPLING OF TWO OSCILLATORS

In this section the theory of weakly coupled oscillators is applied in the analysis of entrainment of two oscillators, as we described formally in section 3.1.2. This subject is interesting in its own right (Cohen *et al.*, 1982; Glass and Mackey, 1979; Storti and Rand, 1986) but also for its use in the analysis of larger systems.

Consider a system of two coupled oscillators satisfying the equations

$$\epsilon \frac{dx_1}{dt} = y_1 - F_1(x_1), \tag{3.3.1a}$$

$$\frac{dy_1}{dt} = -c_1(\delta)x_1 + \delta p_1 \hat{x}_2, \tag{3.3.1b}$$

$$\epsilon \frac{dx_2}{dt} = y_2 - F_2(x_2), \tag{3.3.2a}$$

$$\frac{dy_2}{dt} = -c_2(\delta)x_2 + \delta p_2 \hat{x}_1, \tag{3.3.2b}$$

where $\hat{x}_i(t) = x_i(t-\rho)$, $\rho \geqslant 0$.

In the next two sections we compute the entrained solution for piece-wise linear as well as Van der Pol oscillators coupled without and with delay. In sections 3.3.3 and 3.3.4 we consider two coupled oscillators with different period and limit cycle.

3.3.1 Piece-wise linear oscillators

We consider the case where $c_i(\delta) = 1-\delta q_i$ and

$$F_i(x) = -x \quad \text{for } |x| \leqslant 1, \tag{3.3.3a}$$

$$F_i(x) = x-2 \quad \text{for } x>1, \tag{3.3.3b}$$

$$F_i(x) = -x+2 \quad \text{for } x<-1, \quad i = 1,2. \tag{3.3.3c}$$

For $p_i = q_i = 0$ the period is asymptotically $T_0 = 2\ln 3$. The phase functions are approximated by

$$\phi_i(t) = \alpha_i(1-\delta q_i)t - \delta p_i \int_0^t \frac{X_0(\alpha_{3-i}+\bar{t}-\delta)}{X_0(\alpha_i+\bar{t})} d\bar{t}. \tag{3.3.4}$$

By this expression for the phase an iteration map is defined as follows. Let the phases at time $t = kT_0$ be $\alpha_i^{(k)}$, then

$$\alpha_i^{(k+1)} = \alpha_i^{(k)} - \delta q_i T_0 + p_i \psi(\alpha_i^{(k)} - \alpha_{3-i}^{(k)} - \rho)\delta, \tag{3.3.5a}$$

$$\psi(\beta) = \int_0^{T_0} \frac{X_0(t-\beta)}{X_0(t)} dt. \tag{3.3.5b}$$

The condition for entrainment reads

$$\alpha_1^{(k+1)} - \alpha_2^{(k+1)} = \alpha_1^{(k)} - \alpha_2^{(k)} = \gamma \tag{3.3.6}$$

or

$$(q_2-q_1)T_0 = p_2\psi(-\gamma-\rho) - p_1\psi(\gamma-\rho), \tag{3.3.7}$$

which is identical to condition (iii) of lemma 3.2.6, see (3.2.15). Taking

$$p_2 = q_2 = \rho = 0$$

we have the case of one oscillator forced by an almost identical one. Entrainment is possible if the detuning q_1 is sufficiently small compared with the forcing, see fig.3.3.1.

Let us define the relative phase shift function

$$\chi(\beta;\rho) = p_1\psi(\beta-\rho) - p_2\psi(-\beta-\rho).$$

If we have

$$\chi(\beta\,;\rho) = (q_1 - q_2)T_0 \text{ for some } \beta,$$

an entrained solution exists, see (3.3.7).

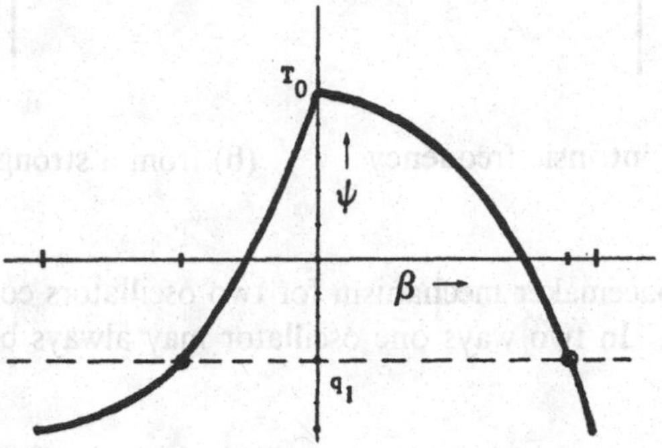

Fig.3.3.1 The phase shift function $\psi(\beta)$. For two phases β the detuning q_1 can be compensated. One of them represents the stable entrained solution.

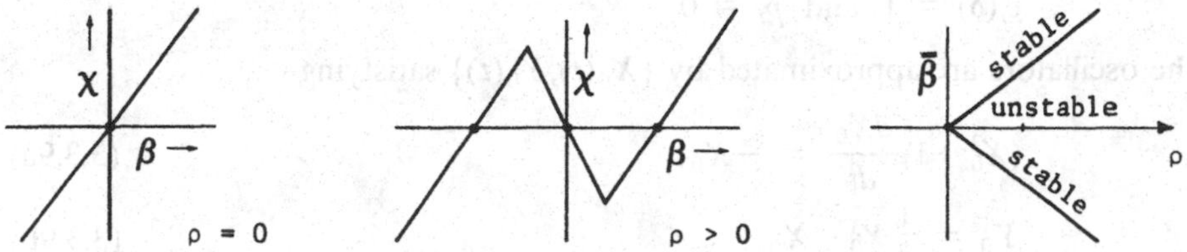

Fig.3.3.2 Dependence of mutual entrainment of identical oscillators upon the delay parameter ρ. For $\rho=0$ the oscillators have the same phase in the entrained state, while for $\rho>0$ either one of the two is advanced in phase.

In fig.3.3.2 it is seen that for two identical oscillators ($q_1 = q_2 = 0$) with equal coupling ($p_1 = p_2 = 1$) the delay acts as a bifurcation parameter. For $\rho = 0$ the entrained solution with $\beta = 0$ is stable. For $\rho>0$ it becomes unstable and two stable solutions with $\beta \neq 0$ branch off. If in a system only one of the two, say oscillator 1, is allowed to be advanced in phase, one has to choose $q_1 < q_2$ or $p_1 > p_2$. Thus, oscillator 1 must have a higher autonomous frequency or a stronger influence upon oscillator 2, see fig.3.3.3.

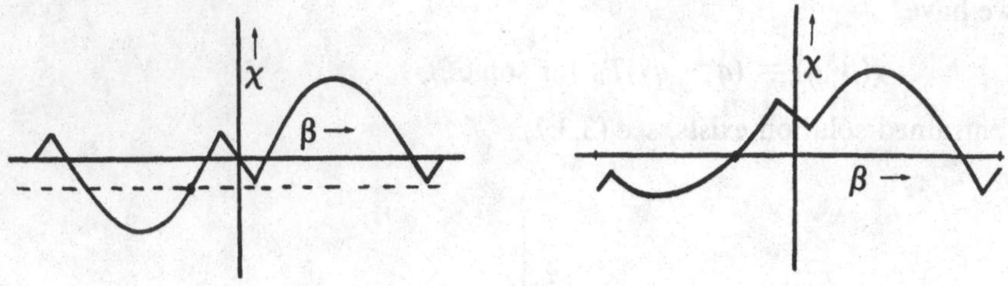

(a) from a higher intrinsic frequency (b) from a stronger forcing

Fig.3.3.3 The pacemaker mechanism for two oscillators coupled with delay. In two ways one oscillator may always be ahead in phase.

3.3.2 Van der Pol oscillators

Next we take in (3.3.1)-(3.3.2)

$$F_i(x_i) = F(x_i) = \frac{1}{3}x_i^3 - x_i. \tag{3.3.8}$$

For

$$c_i(\delta) = 1 \text{ and } p_i = 0$$

the oscillators are approximated by $\{X_0(t), Y_0(t)\}$ satisfying

$$(X_0^2 - 1)\frac{dX_0}{dt} = -X_0, \tag{3.3.9a}$$

$$Y_0 = \frac{1}{3}X_0^3 - X_0. \tag{3.3.9b}$$

Thus, X_0 is implicitly given by

$$\frac{1}{2}X_0^2 - \ln X_0 = -t \quad \text{for} \quad -\frac{1}{2}T_0 < t \leqslant 0, \tag{3.3.10a}$$

where $T_0 = 3 - 2\ln 2$ is the period. Moreover, we have

$$X_0(t) = -X_0(t - \frac{1}{2}T_0) \quad \text{for} \quad 0 < t \leqslant \frac{1}{2}T_0. \tag{3.3.10b}$$

The phase shift function $\psi(\beta)$ given by (3.3.5b) can be evaluated numerically, see fig.3.3.4. This asymptotic expression for the phase shift is compared with numerical results for (3.3.1)-(3.3.2) with ϵ and δ being a fixed small positive values:

$$\epsilon = .0025, \quad \delta = .05. \tag{3.3.11}$$

In the x,y-plane the phase in a point (x,y) is defined as the value of t at the nearest point on the limit cycle (X_0, Y_0) defined by (3.3.10). Numerical integration of (3.3.1)-(3.3.2) with

$$p_1 = 1 \text{ and } p_2 = q_1 = q_2 = \rho = 0$$

for different starting values of the phase difference between the two oscillators yields the phase shifts after one period. In fig.3.3.4 it is seen that the numerical and asymptotic phase shift function agree quite well.

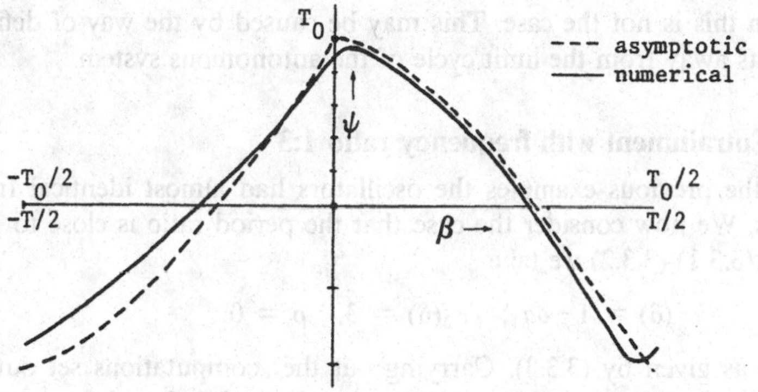

Fig.3.3.4 Phase shift function for the Van der Pol oscillator: a comparison with numerical results

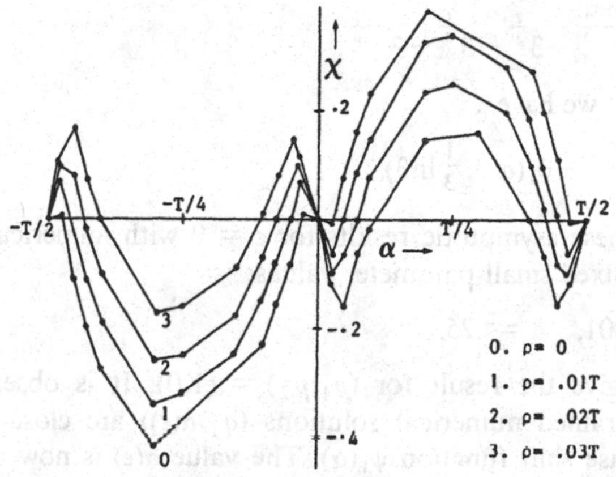

Fig.3.3.5 The relative phase shift function for two Van der Pol oscillators coupled with a delay ρ

In a similar way we may compare the asymptotic solution of two Van der Pol oscillators coupled with delay with numerical solutions of (3.3.1)-(3.3.2) with (3.3.11). We take

$$q_1 = q_2 = 0 \text{ and } p_1 = p_2 = 1.$$

The asymptotic solution is characterized by the relative phase shift function

$$\chi(\alpha) = \frac{1}{2}\psi(\alpha-\rho)-\frac{1}{2}\psi(-\alpha-\rho) \tag{3.3.12}$$

with the function $\psi(\beta)$ similar to (3.3.5b). In fig.3.3.5 we give the numerical equivalent of $Q_1(\alpha)$ for different values of ρ. It is noted that for $\rho = 0$ the discontinuities in the derivatives of ψ at $\alpha = 0$ should cancel. In the numerical solution this is not the case. This may be caused by the way of defining phase in points away from the limit cycle of the autonomous system.

3.3.3 Entrainment with frequency ratio 1:3

In the previous examples the oscillators had almost identical free running periods. We now consider the case that the period ratio is close to 1:3. In the system (3.3.1)-(3.3.2) we take

$$c_1(\delta) = 1-\delta q_1, \quad c_2(\delta) = 3, \quad \rho = 0 \tag{3.3.13}$$

and F_i as given by (3.3.3). Carrying out the computations set out in section 3.2.2 we arrive for $p_i = 1$, $i = 1,2$ at the phase shift functions

$$\psi_2(\alpha) = \psi_1(-\alpha), \quad \alpha = \alpha_1-\alpha_2,$$

$$\psi_1(\alpha) = e^{\alpha}(-\frac{4}{3}-k)+\frac{4}{3}e^{3\alpha} \text{ for } 0\leqslant\alpha<\frac{1}{3}\ln 3$$

with

$$k = 3^{-\frac{1}{3}}+3^{\frac{2}{3}}-3^{\frac{1}{3}}-3^{-\frac{2}{3}}.$$

For $-\frac{1}{3}\ln 3<\alpha\leqslant 0$ we have

$$\psi_1(\alpha) = -\psi_1(\alpha+\frac{1}{3}\ln 3). \tag{3.3.14}$$

Let us compare these asymptotic results for $\epsilon = 0$ with numerical solutions of (3.3.1)-(3.3.2) for fixed small parameter values:

$$\epsilon = .001, \quad \delta = .25.$$

In fig.3.3.6a we give the result for $(p_1,p_2) = (1,0)$. It is observed that the values of the entrained numerical solutions $(q_1,\alpha(\epsilon))$ are close to the stable branch of the phase shift function $\psi_1(\alpha)$. The value $\alpha(\epsilon)$ is now defined as the difference in time at the successive intersections of $x_1(t)$ and $x_2(t)$ with the line $x = 0$. For the case $(p_1,p_2) = (0,1)$ the outcome is quite different, see fig.3.3.6b. The phase shift curve appears to be very sensitive to the value of ϵ. The bandwidth of entrainment is reduced by a factor 2 at $\epsilon = .002$ and again at $\epsilon = .004$. At $\epsilon = .005$ 1:3 entrainment virtually breaks down. Finally, in fig.3.3.6c we sketch the result of mutually entrained numerical solutions: $(p_1,p_2) = (1,1)$. The values $(q_1,\alpha(\epsilon))$ are away from the stable branch of the relative phase shift function

$$\chi(\alpha) = \psi_1(\alpha) - \psi_1(-\alpha). \tag{3.3.15}$$

A further comparision shows that the result is consistent with the values $(q_1,\alpha(\epsilon))$ obtained for $(p_1,p_2) = (0,1),(1,0)$.

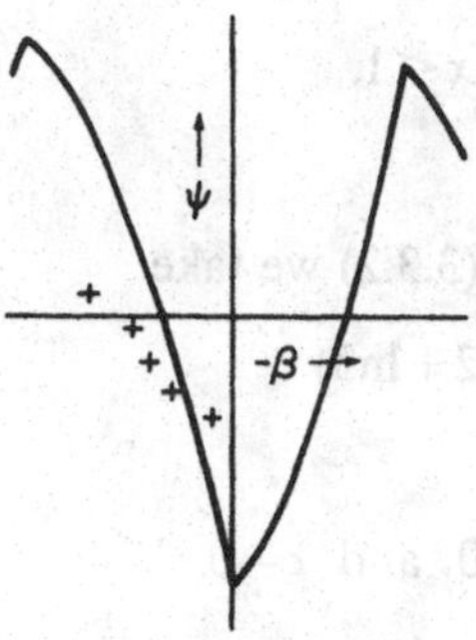

(a) $(p_1,p_2) = (1,0)$

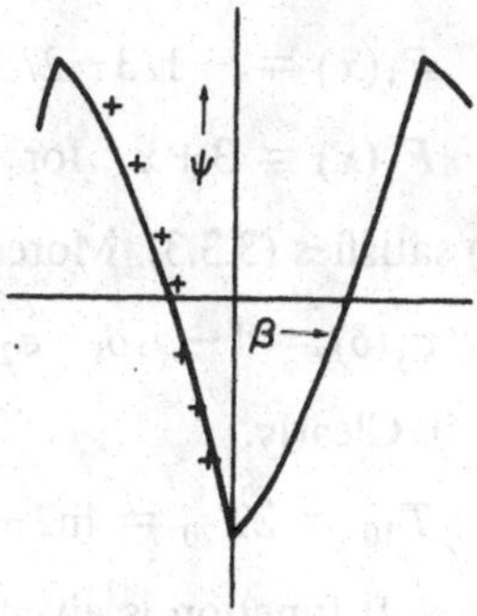

(b) $(p_1,p_2) = (0,1)$

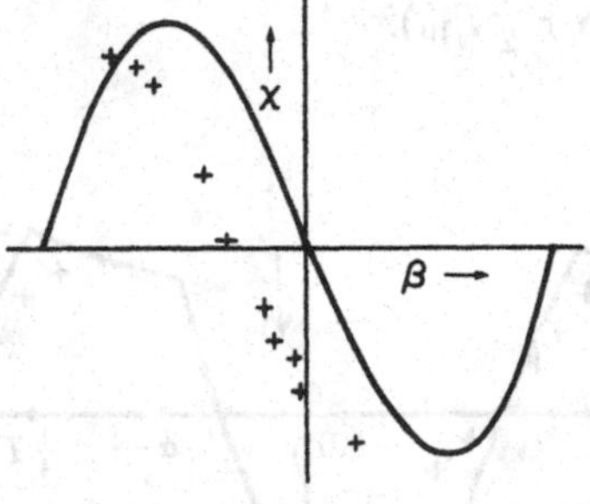

(c) $(p_1,p_2) = (1,1)$

Fig.3.3.6 Numerical $(++)$ and asymptotic solutions of (3.3.1)-(3.3.3) with (3.3.12)

3.3.4 Oscillators with different limit cycles

When carrying out the computations of the foregoing section for autonomous frequencies with a ratio of about 1:2 we find that the phase shift functions ψ_1 and ψ_2 are identically zero in the first order approximation with respect to δ. The computation of the next order term is possible but quite laborious. A further investigation shows that the cancellation is due to symmetry and that it occurs for any ratio containing an even integer with $F_i(x) = -F_i(-x)$. For coupled oscillators that run different, nonsymmetric limit cycles, one can compute the phase shift functions as well, as is seen in the following example:

$$F_1(x) = -2+x \quad \text{for} \quad x \geqslant 1,$$

$$F_1(x) = -1/3-2/3x \quad \text{for} \quad -2<x<1,$$

$$F_1(x) = 3+x \quad \text{for} \quad x \leqslant -2$$

and $F_2(x)$ satisfies (3.3.3). Moreover, in (3.3.1)-(3.3.2) we take

$$c_1(\delta) = 1-q_1\delta, \quad c_2(\delta) = 2\ln 2/(\ln 2+\ln 3)$$

and $p_1 = 0$. Clearly,

$$T_{10} = 2T_{20} = \ln 2+\ln 3 \quad \text{for} \quad \delta = 0 \quad \text{and} \quad \epsilon \to 0.$$

The phase shift function is given by

$$\psi_1(\alpha) = -.63e^{\alpha} - .30e^{c_2\alpha} \quad \text{for} \quad -\tfrac{3}{4}\ln 2+\tfrac{1}{4}\ln 3<\alpha\leqslant 0,$$

$$\psi_1(\alpha) = 8.17e^{\alpha} - 12.86e^{c_2\alpha} \quad \text{for} \quad -\tfrac{1}{4}T_{10}<\alpha\leqslant -\tfrac{3}{4}\ln 2+\tfrac{1}{4}\ln 3,$$

$$\psi_1(\alpha) = -\psi_1(\alpha-\tfrac{1}{4}T_{10}) \quad \text{for} \quad 0<\alpha\leqslant\tfrac{1}{4}T_{10},$$

$$\psi_1(\alpha) = \psi_1(\alpha+\tfrac{1}{2}T_{10}).$$

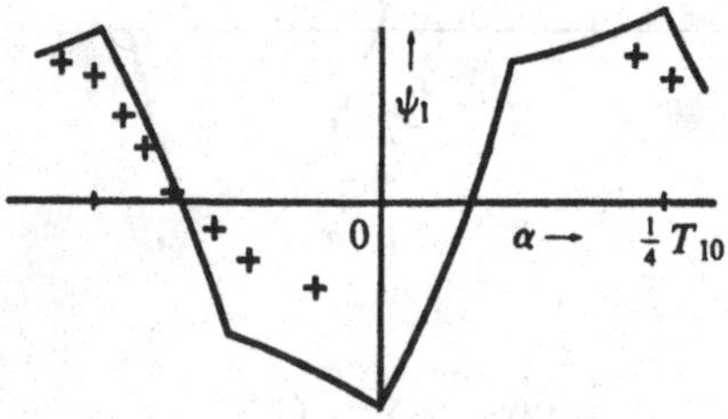

Fig.3.3.7 Phase shift function for frequency ratio 1:2 of two oscillators with different limit cycles

In fig.3.3.7 entrained numerical solutions are plotted for $\epsilon = .001$ and $\delta = .25$. Compared with the case $(p_1,p_2) = (1,0)$ of the preceding example we note a higher sensitivity with respect to ϵ, probably due to the additional discontinuity in the derivative of the phase shift function.

Exercises

3.3.1 How does the choice of the type of coupling in formula (3.3.1) facilitates the computations.
3.3.2 Analyse the two coupled electronic oscillators of Gollub *et al.* (1978), see formula (1.6.1).
3.3.3 Take a piece-wise linear oscillator different from (3.3.3) and compute the phase shift function.
3.3.4 Analyse the bifurcation at $\rho=0$ for the unstable entrained state with the two oscillators of section 3.3.1 in opposite phase, see Mayeri (1973) for a biological application.

3.4 MODELING BIOLOGICAL OSCILLATIONS

From biological point of view it is worthwile to consider a specific type of coupling (Grasman, 1984). Let us assume that the two components x and y represent a biochemical reactant which may diffuse from one compartment of the biological system to the other, see Torre (1975) and Neu (1980). When

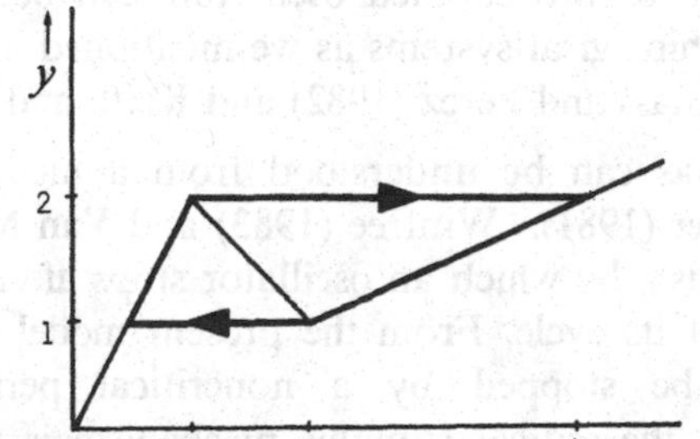

Fig.3.4.1 Limit cycle of a piecewise linear oscillator of type (3.2.1) with $a_i = 1.5$

diffusion of the component x can be neglected, we arrive at a system of type (3.2.1) with

$$h_i(x,y) = p\sum_{j\neq i} a_{ij}(y_j - y_i), \quad p>0, \tag{3.4.1}$$

where a_{ij} is 1 for neighbouring compartments i and j and zero in all other cases. We analyse the dynamics of such a system of coupled oscillators for different types of configurations. It is assumed that each oscillator has the same limit cycle as given in fig.3.4.1. Similar to the examples, given in section 3.3, analytical expressions for the phase shift functions can be computed. However, this is quite laborious and not necessary for obtaining quantitative results on entrainment.

3.4.1 Entrainment with frequency ratio $n:m$

We take two mutually coupled oscillators with

$$c_1(\delta) = 1 \quad \text{and} \quad c_2(\delta) = c. \tag{3.4.2}$$

The period of oscillator 1 is

$$T_0(0) = 2\ln 2 - \frac{1}{2}\ln 5. \tag{3.4.3}$$

The set of algebraic equations (3.1.18) becomes

$$T + \delta\psi_{12}(\beta) = \delta q \text{ (mod) } T_0, \tag{3.4.4a}$$

$$cT + \delta\psi_{21}(-\beta) = \delta q \text{ (mod) } T_0, \tag{3.4.4b}$$

where β denotes the phase difference between the two oscillators in a state of entrainment with $T = mT_0, m = 1,2,...$ In fig.3.4.2 we give the domains in the c,p-plane, where the algebraic equation (3.4.4) or $Q(\alpha) = 0$ of lemma 3.2.6 has a solution for $T = T_0, 2T_0, 3T_0, 4T_0, 6T_0$ and $12T_0$. These solutions come in pairs (a stable and an unstable one). Each domain corresponds with an entrained solution of entrained period ratio $n:m$. Note that the domains overlap and that oscillator 2 can be fixated in two different phases for c small.

This configuration of two coupled oscillators can be seen as a model for $n:m$ entrainment in biological systems as we mentioned in section 3.1, see also Ermentrout (1981), Glass and Perez (1982) and Keith and Rand (1984).

Cardiac arrythmias can be understood from a model with two coupled oscillators, see Keener (1981). Winfree (1983) and Van Meerwijk *et al.* (1983) investigate a mechanism by which an oscillator stops after receiving a stimulus at a critical phase of its cycle. From the present model we conclude that an osillator can also be stopped by a noncritical periodic stimulus. The problematic point in the critical stimulus mechanism is the instability of the phaseless set in the state space. Van Meerwijk *et al.* (1983) observe a tendency of the system to return in the stable oscillatory state, Winfree postulates the presence of a black hole at that spot: an attractor within the phaseless set. In the present model the stopping of a coupled oscillator is a stable mechanism, occurring in two different states. Compared with other entrained solutions at that parameter values they may have a small domain attraction, so that also in this case a stimulus may be needed. In terms of cardiac arrythmias it can be seen as an AV-block with ratio 1:∞. This phenomenon has been noticed too by Ypey *et al.* (1982). In their study the periodic stimulus needs to have a high frequency. Presumably, the domain of attraction of the driven oscillatory state then vanishes and the state of phase fixation takes over.

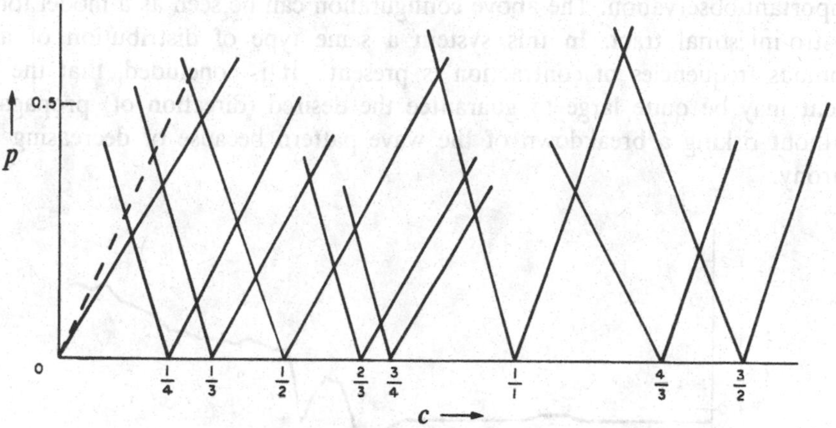

Fig.3.4.2 Domains of $n:m$ entrainment in the parameter plane for two oscillators with widely different intrinsic frequencies

3.4.2 A chain of oscillators with decreasing autonomous frequency

It is assumed that for n oscillators equation (3.4.1) is satisfied for $j = i \pm 1$, $1 \leqslant j \leqslant n$ and that $h_{ij} = 0$ in all other cases. Moreover, we set

$$c_i(\delta) = c_0 + i\Delta c, \quad i = 1,...,n. \tag{3.4.5}$$

Instead of trying to solve the algebraic equation (3.2.20) with T necessarily very large, we compute numerically the iteration map P for $T = T_0$ and carry out a number of iterations until a stable pattern of actual phases and phase velocities arises. The phase velocities are averaged over the last 12 iterations. In the simulation the pattern is independent of the choice of initial phases. About n iterations are needed to arrive at such a pattern. Depending on the gradient of c_i we observe the formation of compartments of oscillators with equal phase velocities, see fig.3.4.3. For a small gradient the compartments, also called plateaus are large, see Ermentrout and Kopell (1983). In the extreme case we have 1 compartment being the complete chain in a fully entrained state. For larger gradients the compartments shrink and increase in number. The size of the compartments is not equally distributed over the chain. In the present situation the one at the slow end dominates. In Grasman and Jansen (1979), where a different type of coupling is analyzed, the one at the fast end is dominating. The present study differs from the ones mentioned above at the point that widely different autonomous frequencies are allowed. In this way we include the effect of large gradients in c_i. From simulations in this range we found that, although the compartments were shrinking and synchrony decreased, the pattern of phase waves, running from the fast to the slow end, persisted. From point of view of biological applications, this is an important observation. The above configuration can be seen as a model for the gastro-intestinal tract. In this system a same type of distribution of autonomous frequencies of contraction is present. It is concluded, that the gradient may be quite large to guarantee the desired (direction of) propagation without risking a breakdown of the wave pattern because of decreasing synchrony.

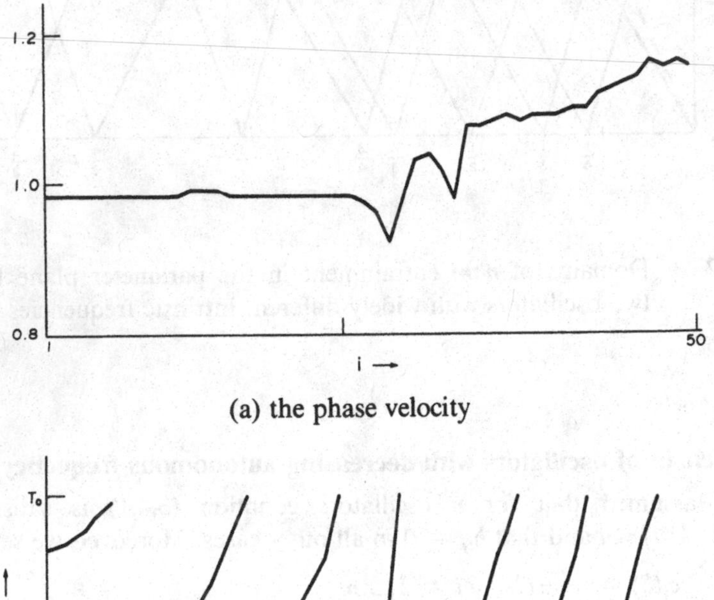

(a) the phase velocity

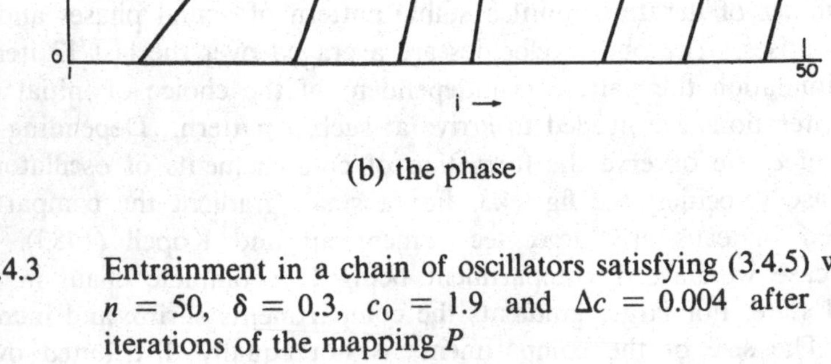

(b) the phase

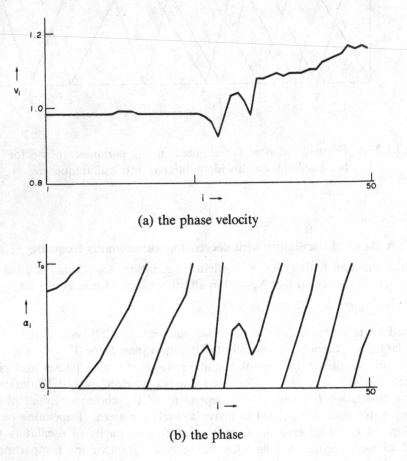

Fig.3.4.3 Entrainment in a chain of oscillators satisfying (3.4.5) with $n = 50$, $\delta = 0.3$, $c_0 = 1.9$ and $\Delta c = 0.004$ after 156 iterations of the mapping P

3.4.3 A large population of coupled oscillators with widely different frequencies

Next we consider a system of n oscillators, all mutually coupled as given by (3.4.1). They are assumed to have autonomous phase velocities that are uniformly distributed over the c_i-interval (0.5,1.5). Again we analyse the iteration map P with the phase velocities averaged over the last 12 iterations. As seen in fig.3.4.4. the spectrum of phase velocities exhibits peaks being spaced in such a way that their relative positions have a ratio $n:m$. Other oscillators move forward and backward over the spectrum without locking in at any of these peaks. There is some analogy with neural oscillators forming a densely coupled system. In theoretical studies of the EEG it is postulated that peaks are due to the mechanism of entrainment. Wiener (1958) speculates that a combination of three peaks could be explained from entrainment of oscillators with nearby frequencies (the central peak). The two side peaks consist of oscillators with frequencies to far away to get entrained. This idea is elaborated in the next section, see also Kreifeldt (1970) and Kuramoto (1975). Our numerical simulations suggest that peaks may occur also as a result of $n:m$ entrainment in case the oscillators have different frequencies. Lopes da Silva *et al.* (1976) observe such peaks in their model of an interacting neuronal population.

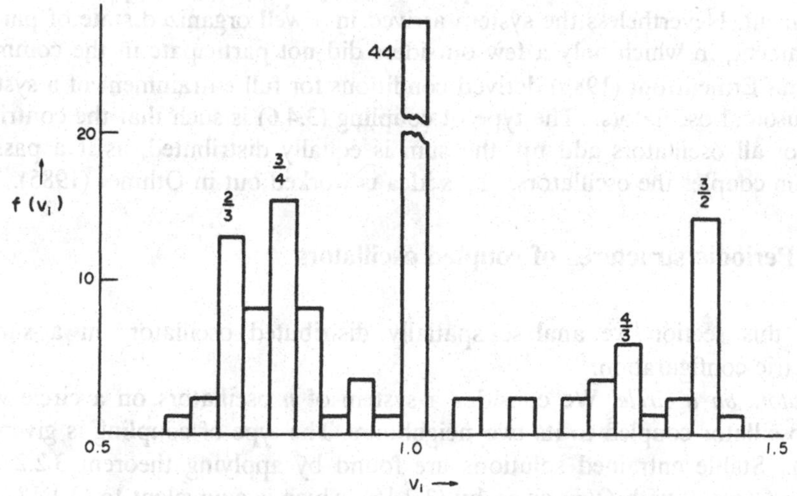

Fig.3.4.4 Distribution of phase velocity in a population of 100 mutually coupled oscillators with intrinsic phase velocity homogeneously distributed over the interval (0.5,1.5)

3.4.4 A large population of coupled oscillators with frequencies having a Gaussian distribution

For this problem we consider a simple type of coupling

$$h_i(x,y) = \sum_{j\neq i} x_j \tag{3.4.6}$$

for the piece-wise linear oscillator given in fig.3.2.2. It is assumed that in

$$c_i(\delta) = 1-q_i\delta \tag{3.4.7}$$

the parameters q_i are drawn independently from a Gaussian distribution with mean zero and standard deviation 2.5. In a numerial experiment we followed the behaviour of 25 oscillators all weakly coupled with $\delta = .02$. The initial phases were drawn independently from a homogeneous distribution over the interval $[0, T_0)$. The actual period of osillator i at the kth iteration is given by

$$T_i^{(k)} = (1+\delta q_i^{(k)})T_0. \tag{3.4.8}$$

In Figure 3.4.5 it is shown how the histogram of the actual periods develops in the course of time. It is also shown how the phases $\alpha_i^{(j)}$ behave as function of q_i. Some remarks have to be added to the text under this figure. In this example the entrained period increases as the number of oscillators increases. It may even happen that the entrained period is outside the range of the free periods. When the range of the free periods is too large a fully entrained state (in which all actual periods are equal) is unattainable. This happened in our experiment. Nevertheless the system arrived in a well organized state of partial entrainment, in which only a few outsiders did not participate in the common rhythm. Ermentrout (1985) derived conditions for full entrainment of a system of sinusoidal oscillators. The type of coupling (3.4.6) is such that the contributions of all oscillators add up, the sum is equally distributed, as if a passive medium couples the oscillators. This idea is worked out in Othmer (1985).

3.4.5 Periodic structures of coupled oscillators

In this section we analyse spatially distributed oscillators in a simple geometric configuration.

Oscillators on a circle. We consider a system of n oscillators on a circle with each oscillator coupled to its two neighbours. The type of coupling is given by (3.4.6). Stable entrained solutions are found by applying theorem 3.2.2 and solving $Q(\alpha)=0$ with $Q(\alpha)$ given by (3.2.15), which is equivalent to (3.1.18):

$$Q_i(\alpha) = \psi_i(\alpha) - \frac{1}{n}\sum_j \psi_j(\alpha) \quad ,i = 1,\ldots,n, \tag{3.4.9}$$

where for the piece-wise linear oscillator of fig.3.2.2

$$\psi_i(\alpha) = Z(\alpha_i - \alpha_{[i-1]}) + Z(\alpha_i - \alpha_{[i+1]}), \quad [j] = j(\text{mod})\, n, \tag{3.4.10a}$$

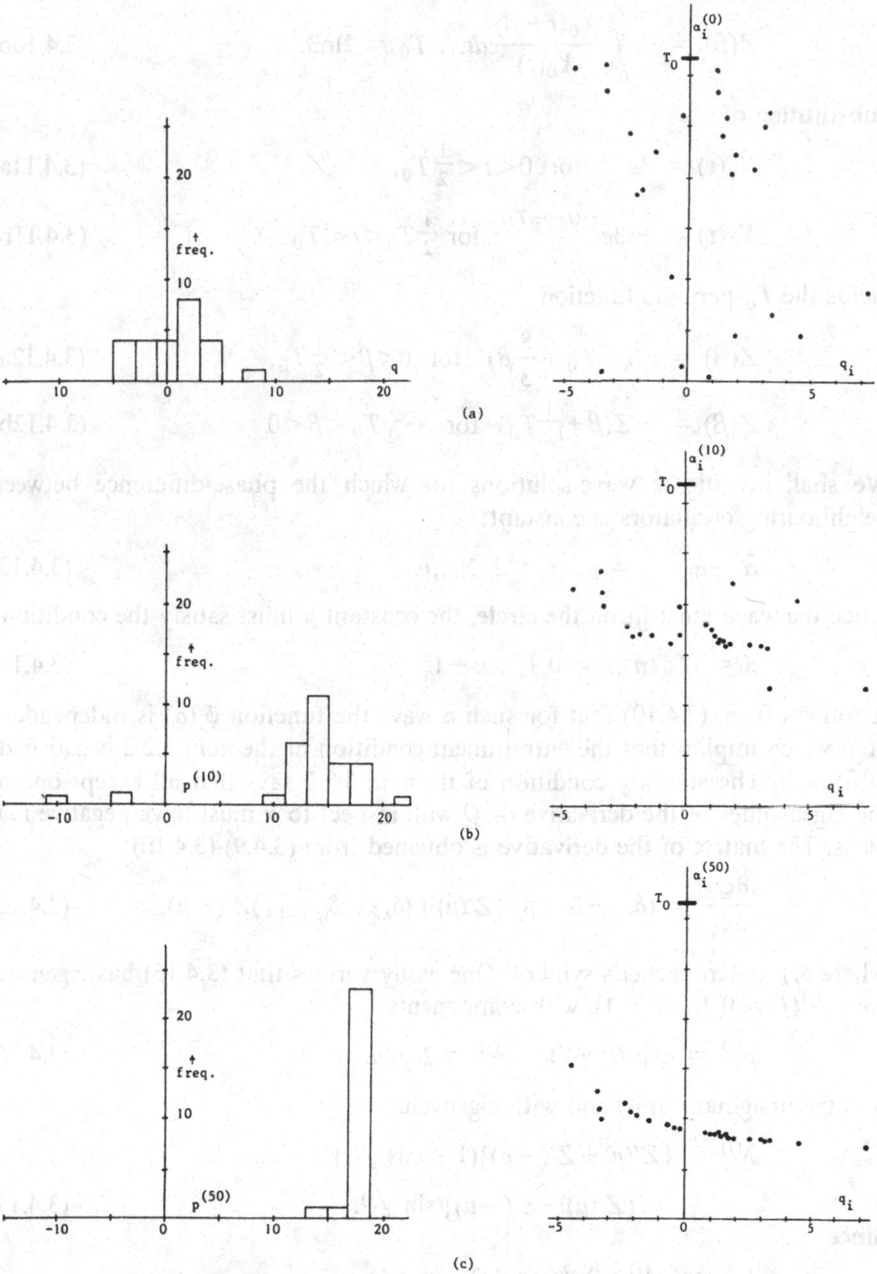

Fig.3.4.5 Development of entrainment for 25 piecewise linear coupled oscillators. (a) describes the situation for the uncoupled oscillators, in (b) and (c) the situation is drawn after 10 resp. 50 iterations of the Poincaré map. On the left the histograms are drawn of the actual periods. On the right the points $(q_i, \alpha_i^{(j)})$ are plotted.

$$Z(\beta) = -\int_0^{T_0} \frac{X_0(t-\beta)}{X_0(t)}dt, \quad T_0 = 2\ln 3. \tag{3.4.10b}$$

Substitution of

$$X_0(t) = 3e^{-t} \text{ for } 0<t<\tfrac{1}{2}T_0, \tag{3.4.11a}$$

$$X_0(t) = -3e^{-(t-\frac{1}{2}T_0)} \text{ for } \tfrac{1}{2}T_0<t<T_0 \tag{3.4.11b}$$

yields the T_0-periodic function

$$Z(\beta) = e^{\beta}(-T_0+\frac{8}{3}\beta) \text{ for } 0<\beta<\tfrac{1}{2}T_0, \tag{3.4.12a}$$

$$Z(\beta) = -Z(\beta+\tfrac{1}{2}T_0) \text{ for } -\tfrac{1}{2}T_0<\beta<0. \tag{3.4.12b}$$

We shall investigate wave-solutions for which the phase-difference between neighbouring oscillators is constant:

$$\tilde{\alpha}_j - \tilde{\alpha}_{[j-1]} = \mu, \ j = 1,2,...,n. \tag{3.4.13}$$

Since the wave must fit on the circle, the constant μ must satisfy the condition

$$\mu = iT_0/n \ ,i = 0,1,...,n-1. \tag{3.4.14}$$

It follows from (3.4.10) that for such a wave the function $\psi_j(\tilde{\alpha})$ is independent of j, which implies that the entrainment condition of theorem 3.2.2 is satisfied: $Q(\tilde{\alpha}) = 0$. The stability condition of theorem 3.2.2 says that all except one of the eigenvalues of the derivative of Q with respect to α must have negative real parts. The matrix of the derivative is obtained from (3.4.9)-(3.4.10):

$$\frac{\partial Q_j}{\partial \alpha_k} = (\delta_{j,k}-\delta_{[j-1],k})Z'(\mu)+(\delta_{j,k}-\delta_{[j+1],k})Z'(-\mu), \tag{3.4.15}$$

where $\delta_{j,k}$ is Kronecker's symbol. One easily verifies that (3.4.15) has eigenvectors $p^{(j)}(j = 0,1,...,n-1)$, with components

$$p_k^{(j)} = \exp(ik\gamma^{(j)}), \quad \gamma^{(j)} = 2\pi j/n, \tag{3.4.16}$$

(i is the imaginary unit) and with eigenvalues

$$\lambda^{(j)} = \{Z'(\mu)+Z'(-\mu)\}(1-\cos\gamma^{(j)}) + i\{Z'(\mu)-Z'(-\mu)\}\sin\gamma^{(j)}. \tag{3.4.17}$$

Since

$$1-\cos\gamma^{(j)}>0 \ (j = 1,2,...,n-1),$$

a sufficient condition for stability is

$$Z'(\mu)+Z'(-\mu)<0. \tag{3.4.18}$$

The period of the entrainment solution is given by

$$T = T_0-\delta[Z(\mu)+Z(-\mu)]+O(\delta^{3/2})+O(\epsilon^{2/3}). \tag{3.4.19}$$

Example. Consider 25 piecewise linear oscillators. Stable waves with $\mu = jT_0/n$ are found for $j = 1,2,...,6$. For $j = 24,23,...,19$ stable waves traveling in the opposite direction are obtained. In the case $j = 0$, in which all oscillators have equal phases (bulk oscillation), the stability condition does not apply, because $Z(\mu)$ is discontinuous in $\mu = 0$. Moreover, the regularity condition of lemma 3.2.6 is not satisfied. This case has been investigated numerically by iteration of the approximate Poincaré mapping $1+\delta Q$. Experiments showed that the bulk oscillation is stable. In calculatons with random initial values the system tended to one of the above stable waves or to the bulk oscillation.

Oscillators on a torus. Let the position of an oscillator on a torus be given by double indices, i,j. Then a system of identical oscillators on a torus with delayed coupling between direct neighbours is described by

$$\epsilon \frac{dx_{j,k}}{dt} = y_{j,k} - F(x_{j,k}) \; (j = 1,2,...,N),$$

$$\frac{dy_{j,k}}{dt} = -x_{j,k} + \delta h_{j,k}(\hat{x}) \; (k = 1,2,...,M),$$

where

$$h_{j,k}(\hat{x}) = \hat{x}_{[j-1,k]} + \hat{x}_{[j+1,k]} + \hat{x}_{[j,k-1]} + \hat{x}_{[j,k-1]} + \hat{x}_{[j,k+1]}, \qquad (3.4.21)$$
$$[j,k] = j(\text{mod})\ N,\ k(\text{mod})\ M.$$

Here $\wedge$ denotes the delay ρ between two neighbouring oscillators, see section 3.2.4. The torus is a simple two dimensional structure without boundary elements. The entrainment properties can be analyzed completely . The analysis and the results are analogous to those for the circle. We restrict ourselves to a statement of the results. An entrained wave solution $\hat{\alpha}$ satifies

$$\tilde{\alpha}_{[j+1,k]} - \tilde{\alpha}_{[j,k]} = \mu = nT_0/N \quad (n = 0,1,...,N-1),$$
$$\tilde{\alpha}_{[j,k+1]} - \tilde{\alpha}_{[j,k]} = \nu = mT_0/\mathrm{M} \quad (m = 0,1,...,M-1). \qquad (3.4.17b)$$

A sufficient condition for stability is

$$Z'(\mu+\rho) + Z'(-\mu+\rho) < 0,$$
$$Z'(\nu+\rho) + Z'(-\nu+\rho) < 0.$$

This means that the stability condition on a torus can be decomposed in two stability conditions on a circle. The period of the wave solution satisfies

$$T_\delta(\epsilon) = T_0 - \delta[Z(\mu+\rho) + Z(-\mu+\rho) + Z(\nu+\rho) + Z(-\nu+\rho)] + O(\delta^{3/2}) + O(\epsilon^{2/3}).$$

Example. We consider 144 piecewise linear oscillators with $M = N = 12$ and $\rho = .002T_0$. The initial phases are drawn independently from a homogeneous distribution on $[0, T_0]$. After 50 iterations of $I+\delta\hat{Q}$ the system reaches a state as sketched in fig.3.4.6, instead of one of the stable waveforms described by (3.4.16)-(3.4.18) irregular waves were running over the torus with wave centres that appeared and disappeared spontaneously. It is not known whether the system would ultimately reach one of the stable waveforms derived above or become periodic with an extremely large period. We only observed that the system persisted in this irregular state for an extended period of time. This phenomenon may explain the uncorrelated contractions of heart cells called fibrillation, see also Strittmatter and Honerkamp (1984).

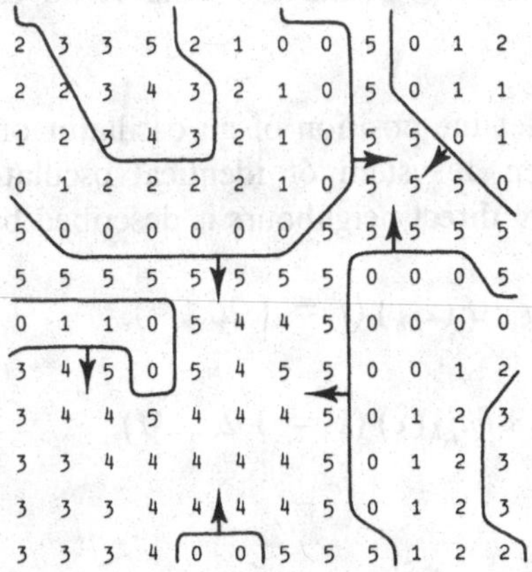

Fig.3.4.6 Piecewise linear oscillators on a torus, illustrating numerical experiment. The right and left boundaries are connected, as well as the upper and lower boundaries. The phase ϕ of each oscillator is indicated by the rounded off value of $6\phi/T_0$. Wave fronts and their direction are indicated by lines and arrows. At the start of the experiment phases were assigned randomly. The figure represents the state after 50 iterations of the Poincaré mapping.

3.4.6 Nonlinear phase diffusion equations

In the beginning of this section on biological oscillations, it was assumed that the biological system has a discrete spatial structure: compartments (cells) with diffusion of a reactant between neighbouring compartments. In a macroscopic approach one may let the size of the compartments tend to zero and consider the problem of oscillations with a continuous spatial dependence leading to a nonlinear diffusion equation. If the solution of this problem is approximated numerically by discretization of the spatial variables, it is better to study it as a system of coupled oscillators as we did in the preceding sections. However, it can be helpful to analyse the diffusion equation in order to explore qualitative properties of such a system, see e.g. Kuramoto (1984). In this way the existence of phase waves of regular or turbulent type can established.

Our method of anaysis of coupled oscillations is easily transmitted to the nonlinear diffusion problem. Let us consider the following Van der Pol type oscillator with diffusion

$$\epsilon\frac{\partial x}{\partial t} = y - F(x), \tag{3.4.20a}$$

$$\frac{\partial y}{\partial t} = -\{1+p(r)\delta\}x + \delta\Delta y, \tag{3.4.20b}$$

where the vector r denotes the spatial dependence and Δ the Laplace operator with respect r. The reference oscillator $\{X_0(t), Y_0(t)\}$ satisfies (3.4.20) for $\delta,\epsilon\to 0$. For only $\epsilon\to 0$ a solution of (3.4.20), written as

$$\{X_0(\phi(t,r)), Y_0(\phi(t,r))\}, \tag{3.4.21}$$

has to satisfy

$$\frac{\partial \phi}{\partial t} = 1 + p(r)\delta + \delta\{\Delta\phi - \frac{(\nabla\phi.\nabla\phi)}{F'(X_0(\phi))}\}, \tag{3.4.22}$$

see (3.1.7)-(3.1.9). In first order approximation with respect to δ we take

$$\phi_0(r,t) = \alpha(r) + t. \tag{3.4.23}$$

Substitution in (3.4.22) yields the second order approximation

$$\phi(r,t) = \alpha(r) + t + \delta\{tp(r) + \Delta\alpha - (\nabla\alpha.\nabla\alpha)\int_0^t \frac{d\bar{t}}{F'(X_0(\alpha(r)+\bar{t})}\}.$$

This expression for the phase generates a mapping of the phase at time $t = kT_0$ to $t = (k+1)T_0$, where T_0 is the period of the reference oscillator:

$$\phi^{(k+1)} = \phi^{(k)} + \delta\{T_0 p(r) + \Delta\phi^{(k)} +$$
$$-I(\nabla\phi^{(k)}.\nabla\phi^{(k)}\}(\text{mod})\ (T_0), \tag{3.4.24a}$$

$$\phi^{(0)} = \alpha(r), \tag{3.4.24b}$$

where

$$I = \int_0^{T_0} \frac{dt}{F'(X_0(\phi^{(k)}+t))}.$$

Since the argument of X_0 is equally shifted over the period for r fixed, the integral is independent of $\phi^{(k)}$. For the Van der Pol oscillator we have

$$\frac{1}{2}I = \int_0^{\frac{1}{2}T_0} \frac{dt}{X_0^2 - 1} = \int_0^1 -\frac{1}{X_0}dX_0 = \ln 2.$$

The limit behaviour of (3.4.24) for $k\to\infty$ can formally be analyzed from a diffusion equation in a slow time scale

$$\frac{\partial\phi}{\partial\tau} = T_0 p(r) + \Delta\phi - I(\nabla\phi.\nabla\phi), \tag{3.4.25a}$$

$$\phi(0) = \alpha(r), \tag{3.4.25b}$$

which gives the slow change of phase in stroboscopic view.

Let the continuously distributed oscillators form a bounded connected set $D \subset \mathbb{R}^N$. Then an entrained solution of period $T_0 + \delta\mu$ is possible, if a constant μ and a function $\alpha(r)$ modulo a constant exist such that

$$\mu + T_0 p(r) + \Delta\alpha - I(\nabla\alpha.\nabla\alpha) = 0 \text{ in } D, \tag{3.4.26a}$$

$$\frac{\partial\alpha}{\partial n} = 0 \text{ on } \partial D, \tag{3.4.26b}$$

where n is the normal to the boundary.

Periodic boundaries. As in section 3.4.5 we consider oscillators on a circle. Now they are continuously distributed and coupled by diffusion. A single wave on a circle satisfies

$$\alpha(\theta) = \frac{T_0\theta}{2\pi} \tag{3.4.27}$$

and the entrained solution has a period

$$T = T_0 + \delta\frac{IT_0^2}{4\pi^2}. \tag{3.4.28}$$

Following the method of section 3.4.5 the same result is found in the limit $n \to \infty$, see also Auchmuty and Nicolis (1976).

Exercises

3.4.1 Consider a chain of n identical oscillators with nearest neighbour coupling and random initial phases. Carry out the iterations of the mapping P and investigate the mechanism of spontaneous pace-maker oscillations (leading centers).

3.4.2 Investigate a chain of n oscillators as given in exercise 1.7 for $n=2$.

3.4.3 Analyse phase waves in $\mathbb{R}$ from (3.4.26a).

3.4.4 Check that indeed the result (3.4.28) is obtained, if one considers n oscillators on a circle with diffusion coupling to the nearest neighbour and $n \to \infty$.

4. THE VAN DER POL OSCILLATOR WITH A SINUSOIDAL FORCING TERM

A Van der Pol oscillator forced by a linear oscillator may have subharmonic as well as chaotic solutions. Asymptotic approximations are constructed and equivalence between solutions and iterates of an interval mapping is established.

4.1 QUALITATIVE METHODS OF ANALYSIS

The problem we are dealing with in this chapter intrigued pure as well as applied mathematicians over the last fifty years. It started with an observation by Van der Pol and Van der Mark on the coexistence of subharmonic solutions of the equation

$$\frac{d^2x}{dt^2}+\nu(x^2-1)\frac{dx}{dt}+x = \nu b(\nu)k \cos kt, \quad \nu>>1 \tag{4.1.1}$$

for certain values of the parameters, see section 1.6.

A first mathematical investigation of this problem was made by Cartwright and Littlewood (1947, 1951), Cartwright (1950, 1952) and Littlewood (1957ab). They discovered the existence of a family of solutions that behave chaotically. Dmitriev (1983) found similar solutions for ν small.

Later on Levinson (1949) considered a simpler version of (4.1.1). It has the form

$$\frac{d^2x}{dt^2} + \Phi(x)\frac{dx}{dt} + x = bp(t), \tag{4.1.2}$$

where

$$\Phi(x) = \text{sign}(x^2-1) \text{ and } p(t) = \text{sign}\,(\sin\frac{2\pi t}{T}). \tag{4.1.3}$$

In this way solutions can be analyzed explicitly by piecing together solutions at different linearity intervals. Smale (1967) introduced the horse-shoe map (see appendix C) as a tool to describe the dynamics of the chaotic solutions constructed by Levinson. Later on it was discovered that in nature chaotic behaviour of dynamical systems is present at a much wider scale, see section 2.6. It led to a new development in the theory of dynamical systems, see Guckenheimer and Holmes (1984).

By changing $\Phi(x)$ and $p(t)$ of (4.1.3) slightly, one can use the theory of diffeomorphisms to analyse solutions. Levi (1980, 1981) followed this approach using differentiable functions $\Phi_\delta(x)$ and $p_\delta(t)$ with the property that

$$\Phi_\delta(x)\to\Phi(x) \text{ and } p_\delta(t)\to p(t) \text{ as } \delta\to 0.$$

Similar to the prototype problem of Nipp for the autonomous problem, see section 2.1.3, one may formulate a variant for forced oscillations, see Habets (1978).

In sections 4.2 and 4.3 we construct asymptotic solutions of (4.1.1) by using singular perturbation techniques. It is assumed that in (4.1.1)

$$b(\nu) = \alpha + \beta\nu^{-1}, \quad 0 \leqslant \alpha \leqslant 2/3. \tag{4.1.4}$$

The case $\alpha > 2/3$ was analyzed by Lloyd (1972). In the next three sections we describe the results of Levi.

4.1.1 Global behaviour and the Poincaré mapping

We write (4.1.2) as

$$\frac{dx}{dt} = \nu\{y - \Phi(x)\}, \tag{4.1.5a}$$

$$\frac{dy}{dt} = \frac{1}{\nu}\{-x + bp(t)\}. \tag{4.1.5b}$$

For ν sufficiently large, it is possible to find an annular trapping region R in the x,y-plane such that the vector field is always directed into R, see fig. 4.1.1. Taking the cross section

$$\Sigma_\theta = \{(x,y,t) | t = \theta(\text{mod})T\} \tag{4.1.6}$$

the mapping $P_\theta : \Sigma_\theta \to \Sigma_\theta$ of (4.1.5) maps R into itself. From numerical experiments it is known that the set

$$A_k = \{P_\theta^k R | 0 \leqslant \theta < T\}$$

tends to a close curve A within R as $k \to \infty$. As the two different stable subharmonics that coexist must lie in A_k, A cannot be a simple closed curve.

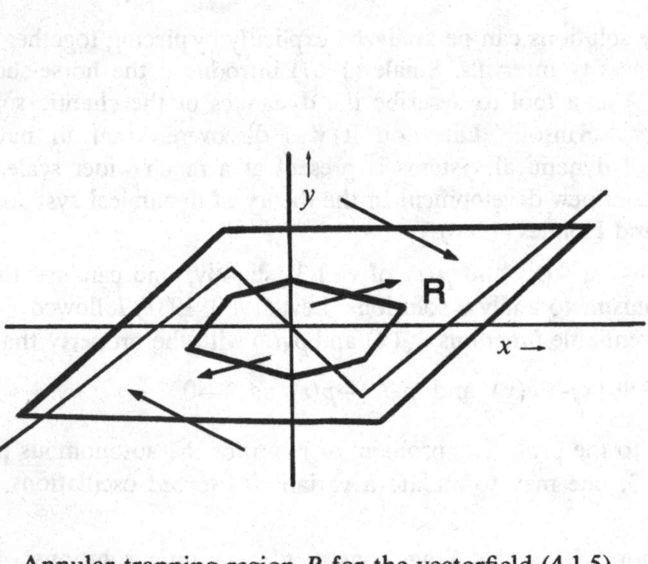

Fig.4.1.1 Annular trapping region R for the vectorfield (4.1.5)

In order to understand this seeming dilemma, we consider the dynamics of (4.1.5) for the set of starting values A_k with k sufficiently large. The set A_k forms a thin annulus. We take a rectangle R^+ within A_k such that for some θ the upper boundary d is mapped into the lower boundary a by P_θ. Consequently, infinitely many iterates of a point of R^+ return in R^+, as P_θ cannot jump over R^+. In fig. 4.1.2 it is visualized how the rectangle R^+ is twisted. It is compressed in one direction and a small section somewhere in the middle is stressed to a length larger than the orginal rectangle R^+. In fig. 4.1.2d it is seen how R^+ is mapped in the rectangle R^-, which is the reflection of R^+ with respect to the origin. The boundaries a and d are mapped in one line that seperates the points which need a different number of iterations to arrive in R^-.

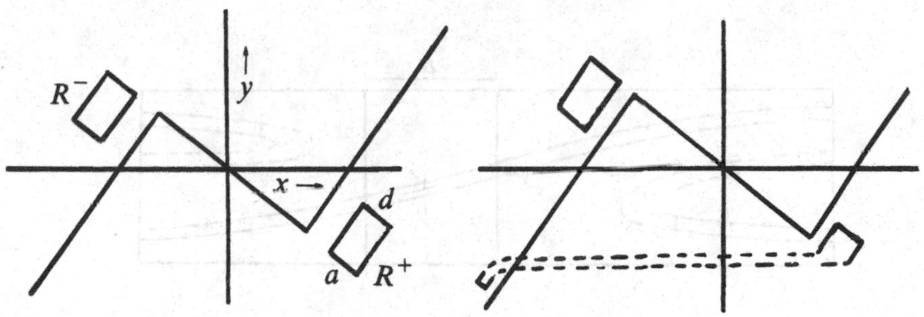

(a) $t = 0$ (b) $t = \frac{1}{2}T$

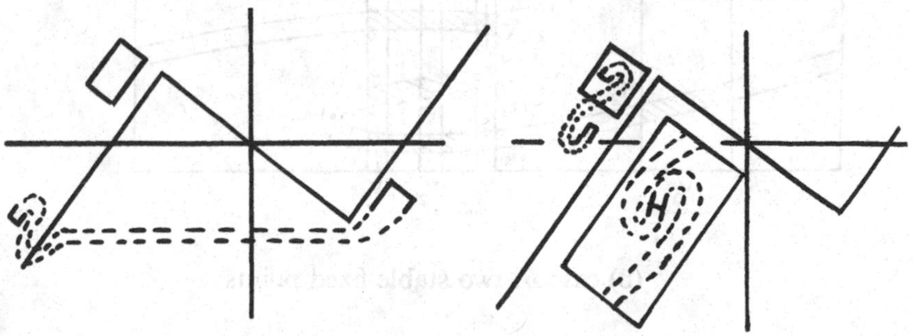

(c) $t = T$ (d) $t = kT$

Fig.4.1.2 Evolution of the rectangle R^+. It is compressed in the direction perpendicular to the stable manifold and stretched during the phase of rapid change. In the inset of (d) the image is completed using $\theta = \theta$ (mod) T.

We now consider the mapping

$$P_\theta : R^+ \rightarrow R^- \tag{4.1.7}$$

in more detail. Actually, the rectangle R^+ is an annulus as the two boundaries a and d can be identified by P_θ. Because of symmetry, we may identify R^- with R^+ and so P_θ is a mapping of an annulus into itself. In fig. 4.1.3a the middle section that is stressed is indicated by Δ. Its image is formed by the solid strip. In this figure we have sketched the case that the mapping has one stable and one unstable fixed point. By rotation we arrive at the more complicated configuration of fig. 4.1.3b: it has two stable fixed points p_1,p_2 and two unstable saddle point type of fixed points z_1 and z_2. There is still another set of points C, which cannot decide to which basin of attractors they belong. C is a Cantor set of measure zero, which is invariant under P. In fact C contains a nontrivial "attractor" of the type known as the horse-shoe.

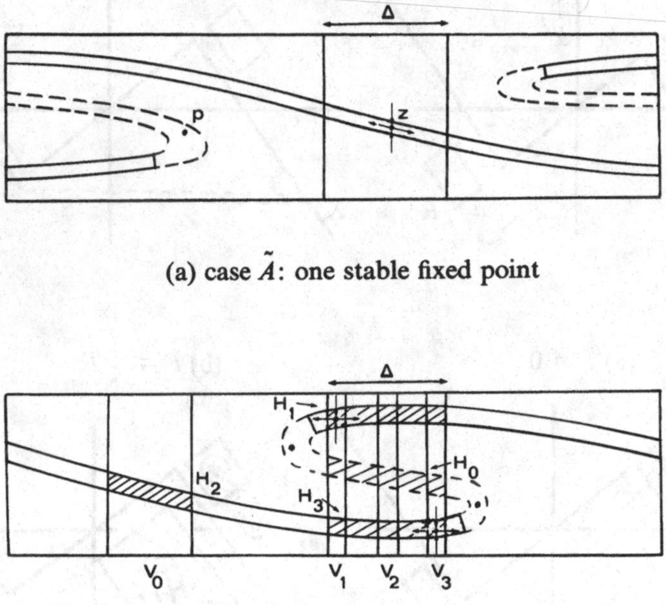

(a) case $\tilde{A}$: one stable fixed point

(b) case $\tilde{B}$: two stable fixed points

Fig.4.1.3 Mapping of the annulus into itself

Symbolic dynamics gives a description of P restricted to C, see Moser (1973). We set

$$\Delta \cap P(\Delta) = \{H_0, H_1, H_3\}, \tag{4.1.8}$$

see fig. 4.1.3b, and define V_0, V_1 and V_3 by

$$P(V_i) = H_i, \quad i = 0,1,3. \tag{4.1.9}$$

Furthermore, let $H_2 = V_0 \cap P(\Delta)$, then V_2 is defined by

$$P(V_2) = H_2. \tag{4.1.10}$$

We now introduce the transition matrix M by

$$M_{ij} = 0, \text{ if } V_i \cap H_j = \emptyset \tag{4.1.11a}$$

and

$$M_{ij} = 1 \text{ otherwise,} \tag{4.1.11b}$$

so

$$M = \begin{bmatrix} 0 & 1 & 1 & 1 \\ 0 & 1 & 1 & 1 \\ 1 & 0 & 0 & 0 \\ 0 & 1 & 1 & 1 \end{bmatrix}. \tag{4.1.11c}$$

4.1.2 The use of symbolic dynamics

The basic idea of symbolic dynamics is to introduce a space Σ of all possible bi-infinite strings of a given set of symbols. Referring to the indices of V_i and H_i of section 4.1.1 we take the symbols 0,1,2 and 3:

$$\Sigma = \{0,1,2,3\}^Z.$$

Thus, an element $a = (..., a_{-1}, a_0, a_1, ...) \in \Sigma$ may read

$$\cdots 01121322110 \cdots.$$

By posing restrictions upon the type of symbol that follows a given symbol one introduces a subspace of Σ:

$$\Sigma_M = \{a \in \Sigma | M_{a_i a_{i+1}} = 1\}.$$

Let M be given by (4.1.11). The mapping $P: C \to C$ is topologically conjugate with the shift $\sigma: \Sigma_m \to \Sigma_m$, where σ satisfies

$$[\sigma(a)]_i = a_{i+1}, \quad i \in \mathbb{Z}.$$

Thus, there is a one to one correspondence between C and Σ_M:

$$\begin{array}{ccc} C & \xrightarrow{P} & C \\ \theta \downarrow & & \downarrow \theta \\ \Sigma_M & \xrightarrow{\sigma} & \Sigma_M \end{array}$$

4.1.3 Some remarks on the annulus mapping

Returning to the annulus mapping we observe the following dependence of this mapping upon b. There is a subdivision of the b-interval into subintervals $\tilde{A}_k$ and B_k separated by small intervals $\tilde{g}_{k_2}$ such that for $b \in \tilde{A}_k$ the mapping P acts as given in fig. 4.1.3a, while for $b \in B_k$ the behaviour can be understood from fig. 4.1.3b. In summary we conclude that for $b \in \tilde{A}_k$ there is only one stable solution and except for one saddle point all solutions tend to this stable one. For $b \in \tilde{B}_k$ two stable solutions exist. Except for a Cantor set of measure zero all solutions tend to one of the attractors.

As b crosses a separation interval $\tilde{g}_k$ there is a sequence of bifurcations and the remarkable phenomenon occurs that for uncountably many $b \in \tilde{g}_k$ there exist infinitely many stable fixed points of P or its iterates, see e.g. Newhouse and Palis (1976). In trying to understand this complicated bifurcation pattern we need more information about the mapping P. Intuitively, it is felt that these bifurcations are related with the disappearance of intersections $V_i \cap H_j$, and that therefore certain finite sequences in Σ_M are forbidden, see also Guckenheimer and Holmes (1984), section 6.7.

It is remarked that the thickness of the annulus is not essential for describing the possible solutions of (4.1.5). One may as well interpret P as a circle mapping from S^1 to S^1. Graphically, P behaves qualitatively as follows (see fig. 4.1.4). A small arc Δ of order $O(\nu^{-1})$ of S^1 is stretched by P, to, say, 1.5 times the length of S^1, while the remaining part $S^1 \setminus \Delta$ is deformed simply by closing the image of Δ. As we will find out, an increase of the forcing b means a clockwise rotation of the image $P(\Delta)$. This immediately leads to the presentation in fig. 4.1.5, where for two different values of b we have given the graph of P. The two cases $\tilde{A}$ and $\tilde{B}$ correspond to the two cases of fig. 4.1.3, see also Alsedà *et al.* (1983).

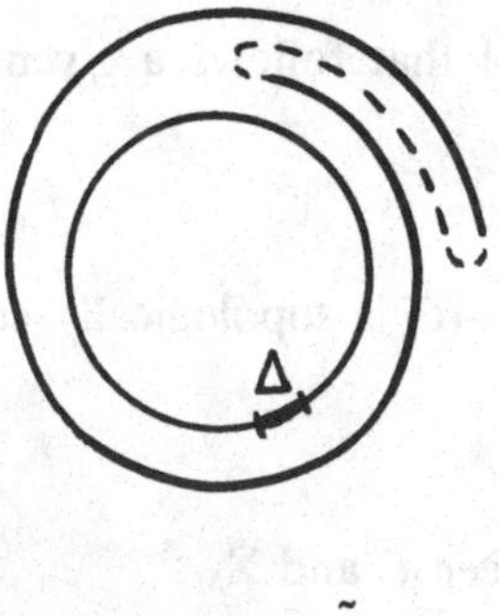

(a) case $\tilde{A}$

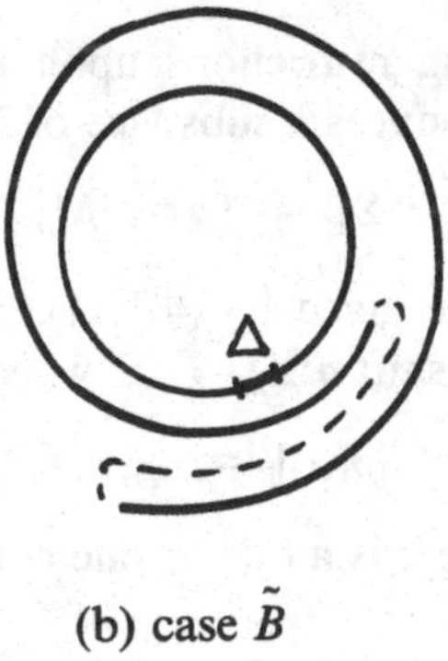

(b) case $\tilde{B}$

Fig.4.1.4 The circle mapping. The image of Δ covers a section with arc length larger than 2π (solid line). The image of S_1/Δ smoothly connects the end points (dotted line).

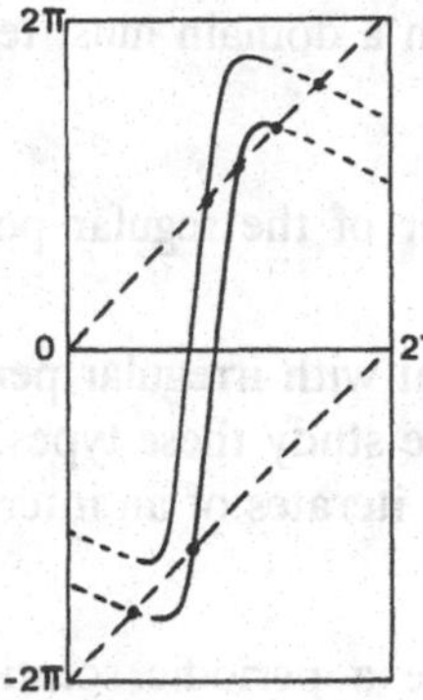

Fig.4.1.5 Graphs of the two cases

It is obvious that an exact description of the mapping P on the interval as sketched in fig. 4.1.5 is the first step in the full understanding of the bifurcation pattern. It is our goal to give a complete description of this mapping and its dependence upon the parameters. In the following sections we will construct matched local asymptotic solutions of (4.1.1), which eventually lead to a mapping on an interval with the same properties as P.

Finally, it is noted that the study of interval mappings has become a field of growing importance in the analysis of dynamical systems, see e.g. Collet and Eckmann (1980). Various bifurcation problems have been solved by using the concept of mapping on an interval, but there still remain many questions about the exact description of families of interval mappings. In particular we mention the chaotic behaviour of mappings within certain parameter ranges.

4.2 ASYMPTOTIC SOLUTION OF THE VAN DER POL EQUATION WITH A MODERATE FORCING TERM

In this section we analyse the forced differential equation

$$\frac{d^2x}{dt^2}+\nu(x^2-1)\frac{dx}{dt}+x = b\cos t\,. \tag{4.2.1}$$

First we construct regular periodic solutions. In the b,ν-plane we determine the domains $\Omega_{m/q}$, where periodic solutions of period $2\pi m$ exist. For

$b \to 0$ the ν-value of points in such a domain must tend to ν_0 satisfying

$$T_0(\nu_0)q = 2\pi m$$

for some q. The rotation number of the regular periodic solution then equals q/m.

Next, in section 4.2.2, we deal with irregular periodic solutions and chaotic solutions. Since in section 4.3 we study these types of solutions in more detail, we just indicate the relation with iterates of an interval map.

4.2.1 Subharmonic solutions

The system (4.2.1) may have a periodic solution with a period being m times the period of the driving term. The conditions on the values of b and ν under which subharmonic entrainment occurs are derived in this section. They are found as the result of a formal approximation of the solution by singular perturbation techniques with $1/\nu$ acting as a small parameter. The conditions bound domains in the b,ν-plane, where a solution with period $2\pi m$ can be constructed formally. It appears that the domains, belonging to different values of m, have overlap. In this section a matched asymptotic approximation of the periodic solution will be made. The method is a generalization of Grasman, Veling and Willems (1976), as it also deals with even subharmonics (Grasman, Jansen and Veling, 1976) and solutions with a fractional rotation number of type q/m.

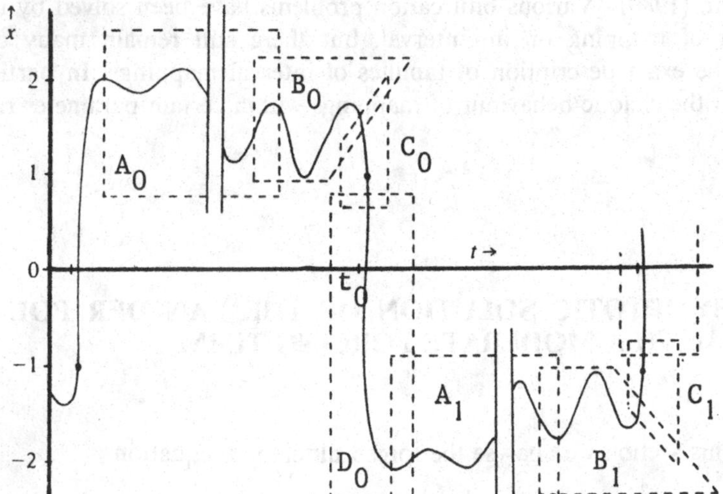

Fig.4.2.1 Regions for local asymptotic solutions of the forced Van der Pol equation (4.2.1)

The entrained solution of (4.2.1) with $b>0$ can be considered as the sum of the autonomous relaxation oscillation and a small harmonic oscillation. We construct local approximations in different regions of the x,t-plane, see fig.4.2.1. In regions of type A a two variable expansion procedure will be applied, see Kevorkian and Cole (1981). In type B and D regions, respectively, the dependent and independent variable are stretched. In C to both variables an asymptotic stretching is applied.

The asymptotic solution for region A_0. For region A_0 we apply the two-variable expansion by introducing another independent variable

$$\tau = (t-t_0)/\nu.$$

It is supposed that the following expansion exists:

$$x = x_0(t,\tau)+x_1(t,\tau)\nu^{-1}+x_2(t,\tau)\nu^{-2}+\cdots. \tag{4.2.2}$$

Substituting this expansion into (4.2.1) and equation terms of $O(\nu)$, we obtain the equation

$$(x_0^2-1)\frac{\partial x_0}{\partial t} = 0. \tag{4.2.3}$$

Hence the function $x_0(t,\tau)$ will only depend on τ. Similarly from terms of $O(1)$, we get the equation

$$(x_0^2-1)(\frac{\partial x_1}{\partial t}+\frac{\partial x_0}{\partial \tau})+x_0 = b\cos t. \tag{4.2.4}$$

This equation contains a slowly varying part

$$\hat{C}_1(\tau)\equiv(x_0^2-1)\frac{\partial x_0}{\partial \tau}+x_0 \tag{4.2.5}$$

which would produce a secular term in the rapidly varying part of x_1:

$$x_1(t,\tau) = \frac{b\sin t}{x_0^2-1} - \frac{\hat{C}_1(\tau)t}{x_0^2-1} + C_1(\tau). \tag{4.2.6}$$

Consequently, $\hat{C}_1(\tau)$ must be taken identically zero, so that

$$\ln x_0 - \tfrac{1}{2}(x_0^2-1) = \tau - D_0, \tag{4.2.7}$$

where D_0 denotes an integration constant. For $x_2(t,\tau)$ the following equation is found:

$$\frac{\partial^2 x_1}{\partial t^2}+(x_0^2-1)(\frac{\partial x_2}{\partial t}+\frac{\partial x_1}{\partial \tau})+ \tag{4.2.8}$$

$$+2x_0x_1(\frac{\partial x_1}{\partial t}+\frac{\partial x_0}{\partial \tau})+x_1 = 0,$$

which has a slowly varying part

$$C_2(\tau) = (x_0^2-1)\frac{\partial C_1}{\partial \tau} + 2x_0 C_1 \frac{\partial x_0}{\partial \tau} + C_1. \tag{4.2.9}$$

In order to remove secular terms in the rapidly varying part of x_2, we also take $C_2(\tau)$ identically zero, so that

$$C_1(\tau) = C_1[x_0(\tau)] = \frac{D_1 x_0}{x_0^2-1}. \tag{4.2.10}$$

The constants D_0 and D_1 may be taken zero, as they denote shifts in time for which already is accounted for by t_0. The functions $x_i(t,\tau)$, $i = 1,2,...$, are singular in $t = t_0$. When t approaches a neighbourhood of magnitude $O(1)$, the first two terms behave as

$$x_0 \approx 1 + \sqrt{t_0 - t}\,\nu^{-1/2}, \quad x_1 \approx \frac{b \sin t}{2\sqrt{t_0 - t}}\nu^{1/2}. \tag{4.2.11}$$

The asymptotic solution for region B_0. In region B_0, where $t = t_0 + O(1)$, the solution will be of the type

$$x = 1 + U(t,\nu)\,\nu^{-1/2}. \tag{4.2.12}$$

Substituting (4.2.12) into (4.2.1) and letting $\nu \to \infty$, we obtain the limiting equation

$$2U_0 \frac{dU_0}{dt} + 1 = b \cos t. \tag{4.2.13}$$

Integration yields

$$U_0(t) = \sqrt{b \sin t + (t_0 - t) + E_0}\,. \tag{4.2.14}$$

For $t_0 - t >> 1$ this solution is expanded as

$$U_0(t) = \sqrt{t_0 - t} + \frac{b \sin t + E_0}{2\sqrt{t_0 - t}} + \cdots, \tag{4.2.15}$$

so that it matches the solution of region A_0 for $E_0 = 0$.

The asymptotic solution for region C_0. Let $t = \bar{t}_0$ be the smallest root satisfying the equation $U_0(t) = 0$ or

$$b \sin \bar{t}_0 + (t_0 - \bar{t}_0) = 0. \tag{4.2.16}$$

In a neighbourhood of this point, where $x \approx 1$ equation (4.2.1) exhibits a turning-point behaviour. We introduce the local coordinate ξ and the new dependent variable $V(\xi)$ by

$$t = \bar{t}_0 + \xi\nu^{-1/3} \quad \text{and} \quad x = 1 + V(\xi)\nu^{-2/3}, \tag{4.2.17}$$

so that the corresponding limiting equation also will contain the term with the second derivative. It reads

$$\frac{d^2 V_0}{d\xi^2} + \frac{dV_0^2}{d\xi} + d = 0, \quad d = 1 - b \cos \bar{t}_0, \tag{4.2.18}$$

and has a general solution of the form

$$V_0(\xi) = \psi'(\xi)/\psi(\xi), \tag{4.2.19}$$
$$\psi(\xi) = \lambda_1 Ai(-d^{-1/3}\xi + d^{-2/3}\mu) + \lambda_2 Bi(-d^{-1/3}\xi + d^{-2/3}\mu).$$

The functions $Ai(z)$ and $Bi(z)$ denote the Airy functions. It appears that $V_0(\xi)$ matches the solution of region B_0 only if $\lambda_2 = \mu = 0$ and $d>0$, as

$$U_0(\bar{t}_0 + \xi\nu^{-1/3}) \approx \sqrt{-d\xi}\,\nu^{-1/6} \quad \text{for} \quad \xi << -1.$$

The condition $d>0$ is equivalent with

$$\cos \bar{t}_0 < 1/b. \tag{4.2.20}$$

The asymptotic solution for region D_0. The Airy function $Ai(z)$ has its largest zero for $z = -\alpha$, so in

$$t = \bar{t}_0 + \xi_0 \nu^{-1/3} \quad \text{with} \quad \xi_0 = \alpha d^{-1/3}$$

the asymptotic solution for region C_0 will be

$$V_0(\xi) \approx (\xi - \xi_0)^{-1}$$

near that point. We introduce a new local variable η according to the transformation

$$t = \bar{t}_0 + \xi_0 \nu^{-\frac{1}{3}} + \eta\nu^{-1}, \tag{4.2.21}$$

so that the first two terms of (4.2.1) become of the same order in magnitude. Substitution in (4.2.1) and multiplication of this equation with ν^{-2} yield the following limiting equation as $\nu \to \infty$:

$$\frac{d^2 W_0}{d\eta^2} + (W_0^2 - 1)\frac{dW_0}{d\eta} = 0. \tag{4.2.22}$$

For the matching with the solution of region C_0 we have the condition

$$W_0(\eta) = 1 + \eta^{-1} \quad \text{as} \quad \eta \to -\infty.$$

This condition is satisfied by selecting solutions of the type

$$\frac{1}{1 - W_0} + \frac{1}{3} \ln \frac{W_0 + 2}{1 - W_0} = -\eta + H_0. \tag{4.2.23}$$

The constant H_0 is found by matching W_0 with higher order terms of the asymptotic solution in region C_0. It turns out that H_0 should depend on $\ln \nu$. However, we will not specify H_0 here as we do not need this result for describing entrainment.

For $\eta >> 1$ we have the estimate

$$W_0(\eta) = -2 + O(e^{-3\eta}). \tag{4.2.24}$$

The asymptotic solution for region A_1. Similar to region A_0 we expand the solution as (4.2.2). The first term satisfies

$$\ln(-x_0)-\frac{1}{2}(x_0^2-1) = \tau+(t_0-t_1)/\nu. \tag{4.2.25}$$

Since the region A_1 starts at $(x,t) = (-2,\bar{t}_0)$, we obtain

$$\bar{t}_1 = \bar{t}_0+(\frac{3}{2}-\ln 2)\nu+\frac{1}{2}b\sin\bar{t}_0+O(\nu^{-1}). \tag{4.2.26}$$

The asymptotic solution for region B_1. We now have

$$x = -1-U_0(t)\nu^{-\frac{1}{2}}+\cdots, \tag{4.2.27a}$$

$$U_0(t) = \sqrt{-b\sin t+t_1-t}, \tag{4.2.27b}$$

so that at time $t = t_1$, satisfying

$$t_1-\bar{t}_1 = b\sin\bar{t}_1, \tag{4.2.28}$$

the solution intersects the line $x = -1$.

The asymptotic solution for region C_1. As for region C_0, an asymptotic solution in terms of Airy functions is found. The condition $d>0$ is translated in

$$\cos\bar{t}_1>-1/b. \tag{4.2.29}$$

Periodicity conditions for rotation numbers $1/m$. The construction of an asymptotic solution for the regions to follow is straight forward and can easily be extracted from the foregoing cases. Let us first consider periodic solutions with period T being a multiple of 2π which intersect the line $x = 0$ twice in a period. Such a solution satisfies

$$\bar{t}_1-\bar{t}_{-1} = 2\pi m, \tag{4.2.30}$$

with $\bar{t}_{-1}$ the time the solution was in $x = -2$ before. According to (4.2.2) this was for

$$\bar{t}_{-1} = t_0-(\frac{3}{2}-\ln 2)\nu+\frac{1}{2}b\sin\bar{t}_{-1}+O(\nu^{-1}). \tag{4.2.31}$$

Let $2\delta(\nu)$ be the difference between the period $T_0(\nu)$ of the autonomous equation and the period T of the special solution, then

$$2\delta = T_0-T = (3-2\ln 2)\nu-2\pi m+O(\nu^{-1/3}). \tag{4.2.32}$$

The system of equations (4.2.16), (4.2.26), (4.2.28), (4.2.30) and (4.2.31) can be reduced to

$$3b(\sin\bar{t}_0-\sin\bar{t}_1) = -4\delta, \tag{4.2.33a}$$

$$b(\sin\bar{t}_0+\sin\bar{t}_1) = -4(\bar{t}_1-\bar{t}_0)+4\pi m. \tag{4.2.33b}$$

It appears that the following change of constants simplifies the calculations

$$\bar{t}_{-1} = 2k_{-1}\pi+\sigma_{-1}$$

$$\bar{t}_0 = (2k_0+1)\pi+\sigma_0 \tag{4.2.34}$$

$$\bar{t}_1 = 2k_1\pi + \sigma_1$$

with $-\pi<\sigma_i\leqslant\pi$, $i = -1,0,1$. In view of the periodicity we have $\sigma_{-1} = \sigma_1$. For $b\leqslant 1$ equations (4.2.16) and (4.2.28) have a unique solution; for $b>1$ we have to select the smallest root. In terms of σ_i the following condition has to be satisfied

$$\sigma_i + b\sin\sigma_i > \sqrt{b^2-1} - \arccos\left(\frac{1}{b}\right) - \pi, \quad i = 0,1. \tag{4.2.35}$$

Conditions (4.2.20) and (4.2.29) transform into

$$\cos\sigma_i > -1/b, \quad i = 0,1. \tag{4.2.36}$$

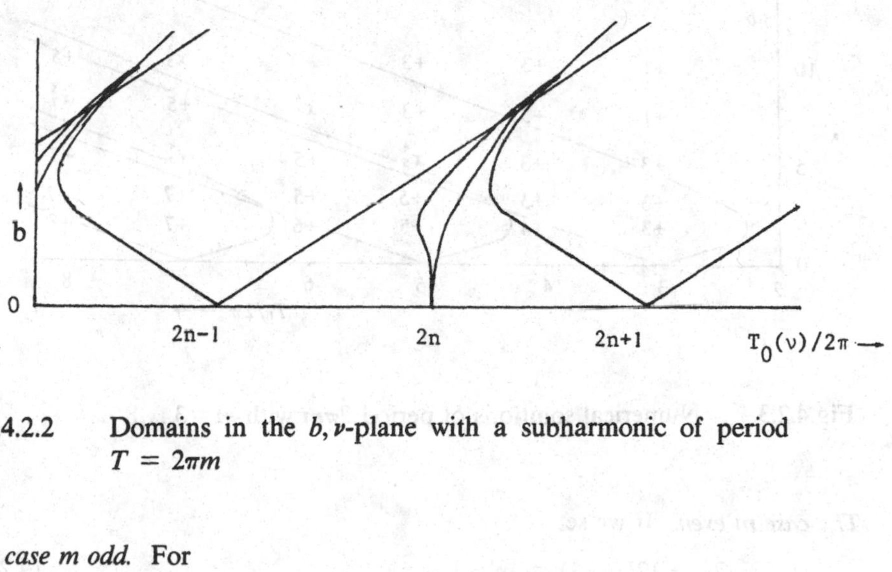

Fig.4.2.2 Domains in the b,ν-plane with a subharmonic of period $T = 2\pi m$

The case m odd. For

$$2k_1-(2k_0+1) = (2k_0+1)-2k_{-1} = m, \tag{4.2.37a}$$

$$\sigma_0 = \sigma_1 \tag{4.2.37b}$$

Eq. (4.2.33b) is satisfied. Substitution in Eq. (4.2.33a) gives

$$\sigma_0 = \sigma_1 = \arcsin\left(\frac{2\delta}{3b}\right), \tag{4.2.38}$$

so another natural restriction of the parameters is

$$\left|\frac{2\delta}{3b}\right| \leqslant 1. \tag{4.2.39}$$

Conditions (4.2.36) are satisfied by (4.2.38). Inequality (4.2.35) reads

$$\arcsin\frac{(2\delta)}{3b} + \frac{2}{3}b > \sqrt{b^2-1} - \arccos\left(\frac{1}{b}\right) - \pi. \tag{4.2.40}$$

In the b, ν-plane (4.2.39) and (4.2.40) determine the region, where a subharmonic solution with period $2\pi m$ with m odd is expected, see fig.4.2.2. These are symmetric solutions satisfying

$$x(t) = -x(t-\tfrac{1}{2}T).$$

The result agrees quite well with values obtained from numerical integration, see fig. 4.2.3.

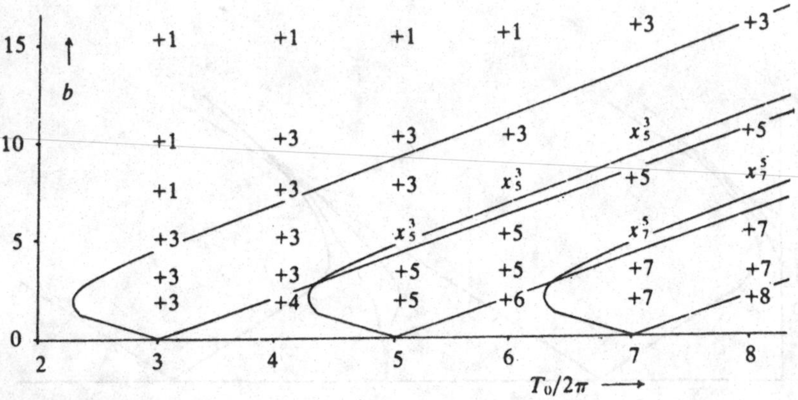

Fig.4.2.3 Numerical solutions of period $2\pi m$ with $m=3,\ldots,8$

The case m even. If we set

$$2k_1-(2k_0+1) = m-1, \tag{4.2.41}$$

$$(2k_0+1)-2k_{-1} = m+1, \tag{4.2.42}$$

then the system (4.2.33) does not admit a solution of the type (4.2.37b). Besides the necessary condition (4.2.39) we obtain, by taking $\sigma_1^+ = \sigma_1+\pi$ and applying the mean value theorem,

$$\frac{\sin^2\sigma_0-\sin^2\sigma_1^+}{\sigma_0-\sigma_1^+} = \frac{16\delta}{3b^2} \tag{4.2.43}$$

the (solvability) condition

$$\left|\frac{16\delta}{3b^2}\right| \leqslant 1. \tag{4.2.44}$$

Finally, we remark that it is possible to give sufficient conditions for solving the system (4.2.33) with m even. These conditions read

$$1-\theta>\frac{4}{b}(\arccos\sqrt{\theta}+\frac{\pi}{2}),\quad \theta=\frac{2\delta}{3b},$$

or

$$1-\theta>\frac{4}{b}\arcsin\sqrt{\theta}.$$

Periodicity with rotation number q/m. It is also possible to construct subharmonic solutions that intersect the t-axis $2q$ times ($q=2,3,...$) in one period,

$$T=(3\dot{-}2\ \ln\ 2)q\nu+O(1).$$

This would lead to a system of $2q$ equations of the type (4.2.33). Such system can easily be reduced to a system of q equations in case of symmetry with

$$x(t)=-x(t-\frac{1}{2}T).$$

4.2.2 Dips, slices and chaotic solutions

The asymptotic solution we constructed in the preceding section is not valid for d close to zero in (4.2.18). Let us first consider the solution $\hat{x}(t;\nu)$ for which exactly

$$d=1-b\cos\bar{t}_0=0. \tag{4.2.45}$$

For $t\leqslant\bar{t}_0$ this solution is represented by (4.2.12)-(4.2.14). For $t\geqslant\bar{t}_0$ the leading term is

$$U_0(t)=-\sqrt{b\sin t+t_0-t}. \tag{4.2.46}$$

As t approaches $\bar{t}_0$, satisfying (4.2.16), we have

$$\hat{x}(t;\nu)=1+O(\bar{t}_0-t) \tag{4.2.47}$$

instead of $1+O(\sqrt{\bar{t}_0-t})$ for $d>0$. For $t\in(\bar{t}_0,\tilde{t}_0)$ the solution $\hat{x}(t;\nu)$ is in the unstable region $|x|<1$ until it again reaches the line $x=1$ at $t=t_0$, see fig. 4.2.4. Next this solution is perturbed such that (4.2.45) is still valid, while (4.2.16) has an exponentially small remainder term $\bar{\delta}(\nu)$ of order $O(e^{-a\nu})$:

$$b\sin\bar{t}_0+t_0-\bar{t}_0=-\sigma\bar{\delta}(\nu),\quad \sigma=\pm1. \tag{4.2.48}$$

This is equivalent with taking $\bar{\delta}(\nu)=0$ and

$$d=\begin{cases}\delta(\nu) & \text{for } \sigma=1,\\ \bar{d}-\delta(\nu) & \text{for } \sigma=-1,\end{cases} \tag{4.2.49}$$

where $\bar{d}$ is the derivative of

$$D(t)=t-b\sin t$$

at the third root of $D(t)=0$ in case of coalescence of the two other ones.

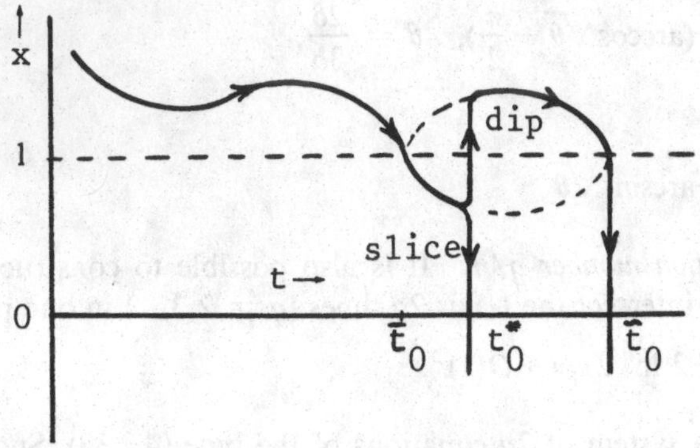

Fig.4.2.4 Dips and slices

Substitution of

$$x(t;\nu) = \hat{x}(t;\nu) + V(t;\nu) \tag{4.2.50}$$

in (4.2.1) yields in first order approximation

$$\frac{d^2V}{dt^2} + \nu\frac{d}{dt}\{(\hat{x}^2 - 1)V\} = 0. \tag{4.2.51}$$

From (4.2.14) and (4.2.48) it is seen that as t approaches a $\nu^{-1/2}$-neighbourhood of $\bar{t}_0$, the following behaviour of the solution is found

$$x(t;\nu) \approx \hat{x}(t,\nu) - \frac{\sigma\nu^{-1/2}\bar{\delta}(\nu)}{2p^2(\bar{t}_0 - t)} \tag{4.2.52}$$

with

$$p^2 = \sqrt{-\frac{b}{2}\sin\bar{t}_0}.$$

Consequently, the perturbation $V(t,\nu)$ must satisfy

$$V(\bar{t}_0 + \xi\nu^{-1/2};\nu) = \frac{\sigma e^{-a\nu}}{2p^2\xi} \quad \text{for } \xi \to -\infty. \tag{4.2.53}$$

The solution of (4.2.51) that meets this condition is

$$V = \sigma e^{-\nu\{A(t)+a\}}\{\frac{\sqrt{\pi}}{2p} - \int_{t_0}^{t} e^{\nu A(\bar{t})}d\bar{t}\}, \tag{4.2.54}$$

$$A(t) = \int_{t_0}^{t}(\hat{x}^2(\bar{t};0) - 1)\, d\bar{t}. \tag{4.2.55}$$

Let a point $t_0^* \in (\bar{t}_0, \tilde{t}_0)$ exist such that

$$A(t_0^*) = -a. \tag{4.2.56}$$

Then at this point the perturbation explodes and the solution enters a boundary layer region of type C. It either tends rapidly to the branch (4.2.13) with $x>1$ (completing a dip for $\sigma = -1$) and follows this branch until $\tilde{t}_0$, where it crosses the unstable region $|x|<1$, or it directly crosses this region (making a slice for $\sigma = 1$) to go immediately to the region A_1. The solution for the region A_1 is given by (4.2.25)-(4.2.26) with $\bar{t}_0$ replaced by either t_0^* (slice) or $\tilde{t}_0$ (dip). The intersection with the line $x = -1$ is at $t = \bar{t}_1$ satisfying (4.2.28).

In this way we have established a relation between $\bar{t}_0$ and $\bar{t}_1$. This is generalized to

$$\bar{t}_k = F(\bar{t}_{k-1}),$$

which gives a complete description of all possible solutions, the subharmonics as well as the chaotic ones. In section 4.3 we explore in more detail the relation between solutions of the differential equation (4.1.1) and the iterates of a mapping.

More in line with the theory of iterated maps on an interval, we many construct a mapping on a compact interval. The natural choice for the variable is

$$d_k = 1 - b \cos \bar{t}_k$$

on the interval $[0, \bar{d}]$. The mapping

$$d_{k+1} = P(d_k; b, \nu)$$

can be constructed explicitly from the formula's given above. It is noted that near $d = 0$ and $d = \bar{d}$ representation (4.2.49) is used. In case the mapping has two stable fixed points, chaotic solutions also exist, as we will see in the next section.

4.3 ASYMPTOTIC SOLUTION OF THE VAN DER POL EQUATION WITH A LARGE FORCING TERM

In this section we study the Van der Pol equation with a large sinusoidal forcing term

$$\frac{d^2x}{dt^2} + \nu(x^2-1)\frac{dx}{dt} + x = (\alpha\nu + \beta)\cos kt, \; \nu >> 1, \tag{4.3.1}$$

with $0<\alpha<2/3$ (Grasman *et al.,* 1984). A numerical investigation is found in Flaherty and Hoppensteadt (1978) and El-Abbasy and James (1983).

4.3.1 Subharmonic solutions

Using singular perturbation techniques we construct a formal asymptotic approximation of the $2\pi(2n-1)$-periodic solution of (4.3.1) with $n = O(\nu)$. In the process of constructing such approximation we arrive at a set of conditions for α, β and ν. These conditions are such that for a given α, the parameter β lies in an interval

$$\underline{\beta}_n(\alpha)<\beta<\overline{\beta}_n(\alpha). \tag{4.3.2}$$

These intervals overlap, so that for β in the interval $(\beta_n(\alpha),\overline{\beta}_{n+1}(\alpha))$ two solutions with period $T = 2\pi(2n\pm1)/k$ coexist. In section 4.2.1 asymptotic solutions for the case $\alpha = 0$ and $k = 1$ have been constructed. In our analysis of the present problem we see that in the asymptotic solution elements of this case can be distinguished, such as the two variable expansion. However, the 2π-periodic oscillations are now present in the first order aproximation, so that the solution cannot be seen as an autonomous relaxation oscillation with an additive small 2π-periodic oscillatory term. Fortunately, an explicit expression for the first order approximation can be derived. Another aspect is the way in which the solution approaches the lines $|x| = 1$: it tends to be tangent to these lines. As a result of this, local aproximations contain parabolic cylinder functions instead of Airy functions. Finally, the solution returns a large number of times in the stable region $|x|>1$ after visiting an $\nu^{-1/2}$-neighbourhood of the line $|x| = 1$, before it definitively crosses the unstable region $|x|<1$. Consequently, a large number of local approximations have to be made, see fig.4.3.1.

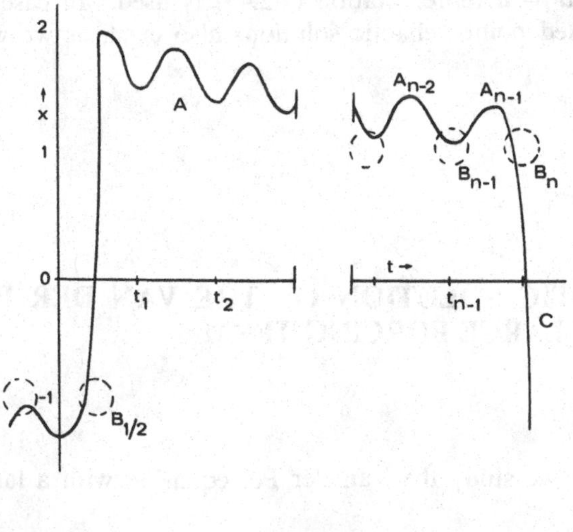

Fig.4.3.1 Regions for local asymptotic solutions of the forced Van der Pol equation (4.3.1)

The regions A_m and B_m. Let us first analyse the local asymptotic solution for a region A_m above the t-axis with m a positive integer, see fig.4.3.1. It is assumed that the solution can be expanded as

$$x(t,\nu) = x_{m0}(t) + \nu^{-1}x_{m1}(t) + \cdots. \tag{4.3.3}$$

Substitution in (4.3.1) yields after grouping of terms of order $O(\nu)$ and $O(1)$ the following equations for x_{mi}, $i = 1,2$,

$$(x_0^2 - 1)\frac{dx_m0}{dt} = \alpha k \cos kt, \tag{4.3.4}$$

$$(x_{m0}^2 - 1)\frac{dx_{m1}}{dt} + 2x_{m0}\frac{dx_{m0}}{dt}x_1 = - \frac{d^2x_{m0}}{dt^2} - x_{m0} + \beta k \cos kt, \tag{4.3.5}$$

or after integration

$$\frac{1}{3}x_{m0}^3 - x_{m0} = \alpha \sin kt + C_0^{(m)}, \tag{4.3.6}$$

$$(x_{m0}^2 - 1)x_{m1} = -\frac{dx_{m0}}{dt} - \int_{t_{m-1}}^{t} x_{m0}(\bar{t})d\bar{t} + \beta \sin kt + C_1^{(m)}. \tag{4.3.7}$$

Since for $t \uparrow t_m$ x approaches the value 1, we have $C_0^{(m)} = \alpha - 2/3$ and asymptotically

$$x \approx 1 - \frac{1}{2}a^2(t - t_m) + K_m\{a^2(t - t_m)\nu\}^{-1} \tag{4.3.8}$$

with

$$K_m = \frac{1}{2}a^2 + (-C_1^{(m)} + \beta + 2\pi I),\ a = \sqrt{2\alpha k^2}, \tag{4.3.9}$$

$$I = \frac{1}{2\pi}\int_{t_{m-1}}^{t_m} x_{m0}(t)dt, \tag{4.3.10}$$

$$x_{m0}(t) = 2\cos[\frac{1}{3}\arccos\{\frac{3}{2}\alpha \sin kt + \frac{3}{2}\alpha - 1) + \frac{2}{3}\pi j\}], \tag{4.3.11}$$

where $j = 0$.

Next we deal with the region B_m. From (4.3.8) we conclude that one may expect a different asymptotic behaviour of the solution in a $\nu^{-1/2}$-neighbourhood of $(x,t) = (1,t_m)$. We, therefore, introduce the local variable

$$\xi = (t - t_m)\nu^{-\frac{1}{2}}$$

and expand x as follows

$$x = 1 + \nu^{-\frac{1}{2}}V_{m0}(\xi) + \nu^{-1}V_{m1}(\xi) + \cdots. \tag{4.3.12}$$

Substitution in (4.3.1) yields for the leading terms of $O(\nu^{1/2})$

$$\frac{d^2 V_{m0}}{d\xi^2}+2V_{m0}\frac{dV_{m0}}{d\xi} = \alpha k^2 \xi. \tag{4.3.13}$$

Furthermore V_{m0} should match (4.3.8) or

$$V_{m0}(\xi)\approx -\frac{1}{2}a^2\xi + K_m(a^2\xi)^{-1} \quad as \quad \xi\to -\infty. \tag{4.3.14}$$

After carrying out a transformation of type $V_{m0} = z'/z$, we find the following solution for (4.3.13) and (4.3.14):

$$V_{m0}(\xi) = -aD'_p(-a\xi)/D_p(-a\xi), \quad p = K_m/a^2, \tag{4.3.15}$$

where $D_p(x)$ denotes the parabolic cylinder function of order p.

Let us assume that K_m is negative, then the parabolic cylinder function has no zeroes and V_{m0} remains regular. The asymptotic solution will loose its validity for $\xi\to\infty$, as

$$V_{m1}(\xi)\approx\frac{1}{2}a^2\xi-(K_m/a^2+1)\xi^{-1}. \tag{4.3.16}$$

Now the solution enters the region A_{m+1}, where (4.3.3)-(4.3.8) hold with m replaced by $m+1$. For $t\downarrow t_m$ this solution behaves as

$$x_{m+1}(t)\approx 1+\frac{1}{2}a^2(t-t_m)-K_{m+1}\{a^2(t-t_m)\nu\}^{-1}. \tag{4.3.17}$$

Consequently, the matching conditions result in the recursive relation

$$K_{m+1} = Q(K_m)\equiv K_m+2\pi I. \tag{4.3.18}$$

Similar matched local solutions are found for the regions A_m and B_m below the t-axis with $m = 1/2, -1/2, -3/2, \ldots$. For such regions A_m (4.3.11) holds with $j = 1$. Clearly, there exists a value of m for which $0<K_m<2\pi I$. Without loss of generality this may be for $m = 1/2$ below the t-axis, see fig.4.3.1. The same occurs above the t-axis the first time for $m = n\geqslant 1$ with n integer.

For $K_n>0$ the parabolic cylinder function has one or more zeros. Let ξ_0 be the first zero one meets coming from $-\infty$. For $\xi\uparrow\xi_0$ we have

$$V_n(\xi)\approx(\xi-\xi_0)^{-1}+\frac{1}{3}a^2(\frac{1}{4}a^2\xi_0^2-K_n-\frac{1}{2})(\xi-\xi_0) \tag{4.3.19}$$

and the solution leaves there the $\nu^{-1/2}$-neighbourhood of $(x,t) = (1,t_n)$. A similar behaviour is found in the region $B_{1/2}$.

There is a strong similarity between the mapping $Q(k)$ of (4.3.18) and $D(x,y)$ of (4.3.8). As explained in section 4.1.3 Levi derived from D a mapping $P:R\to R'$. where R is a strip in the phase plane and R' its reflection with respect to the origin. Likewise we will construct a mapping $P(K)$ of the interval $[0,2\pi I]$ into itself. Starting in region $B_{1/2}$ with $K_{1/2}$ within this interval we look for the smallest integer $n\geqslant 1$ such that K_n returns in this interval, then $P(K_{1/2}) = K_n$. In order to derive an explicit expression for $P(K)$ we have to construct matched local solutions for the remaining regions C and A.

The region C. At $t = t_{1/2} = \xi_0 \nu^{-1/2}$ the solution leaves the $B_{1/2}$-region and enters the boundary layer region C. We introduce the local variable

$$\eta = (t - t_{1/2} - \xi_0 \nu^{-1/2})\nu. \tag{4.3.20}$$

Assuming that the solution can be expanded as

$$x(t,\nu) = w_0(\eta) + \nu^{-1} w_1(\eta) + \nu^{-2} w_2(\eta) + \cdots \tag{4.3.21}$$

we arrive at a recurrent system of equations for w_i:

$$\frac{d^2 w_0}{d\eta^2} + (w_0^2 - 1)\frac{dw_0}{d\eta} = 0 \tag{4.3.22}$$

$$\frac{d^2 w_1}{d\eta^2} + (w_0^2 - 1)\frac{dw_1}{d\eta} + 2w_0 w_1 \frac{dw_0}{d\eta} = 0, \ldots \tag{4.3.23}$$

According to (4.3.19), the local solution (4.3.21) matches the asymptotic solution for region $B_{1/2}$, if for $\eta \to -\infty$

$$w_0 = -1 - 1/\eta, \quad w_1 = -\tfrac{1}{3}a^2(\tfrac{1}{4}a^2\xi_0^2 - K_{1/2} - \tfrac{1}{2})\eta. \tag{4.3.24ab}$$

Condition (4.3.24a) is satisfied by solutions of (4.3.22) of the form

$$\frac{1}{-1 - w_0} + \tfrac{1}{3}\ln\{\frac{w_0 - 2}{-1 - w_0}\} = -\eta + \text{constant}, \tag{4.3.25}$$

while because of (4.3.24b) the integrated equation (4.3.23) must have the form

$$\frac{dw_1}{d\eta} + (w_0^2 - 1)w_1 = -a^2(\tfrac{1}{4}a^2\xi_0^2 - K_{1/2} - \tfrac{1}{2}). \tag{4.3.26}$$

As $\eta \to \infty$ the boundary layer region is left at exponential rate

$$x = 2 - \tfrac{1}{3}a^2(\tfrac{1}{4}a^2\xi_0^2 - K_{\frac{1}{2}} - \tfrac{1}{2})\nu^{-1} + \tag{4.3.27}$$
$$+ O(\exp\{-3(t - t_{1/2}\xi\nu^{-1/2})\nu\}).$$

The two variable expansion for region A. In the region A the solution has an oscillatory behaviour in the $O(1)$ time scale and its average value decreases slowly in the $O(\nu)$ scale. Using the two variable procedure, see Kevorkian and Cole (1981), we write the solution for the region A above the t-axis as

$$x = x_0(t,\tau) + \nu^{-1}x_1(t,\tau) + \nu^{-2}x_2(t,\tau) + \cdots, \tag{4.3.28}$$

with $\tau = (t - t_{1/2})/\nu$. The coefficients satisfy a recurrent system of differential equations. The first one is

$$(x_0^2 - 1)\frac{dx_0}{dt} = \alpha k \cos kt, \tag{4.3.29}$$

or

$$\tfrac{1}{3}x_0^3 - x_0 = \alpha \sin kt + C_0(\tau) \tag{4.3.30}$$

with a solution above the line $x = 1$:

$$x_0(t,\tau) = 2\cos[\frac{1}{3}\arccos\{\frac{3}{2}\alpha\sin kt + \frac{3}{2}C_0(\tau)\}]. \quad (4.3.31)$$

For the second equation we obtain

$$\frac{\partial^2 x_0}{\partial t^2} + (x_0^2-1)(\frac{\partial x_0}{\partial \tau} + \frac{\partial x_1}{\partial t}) + 2x_0x_1\frac{\partial x_0}{\partial t} + x_0 = \beta k\cos kt \quad (4.3.32)$$

or

$$(x_0^2-1)x_1 = -\frac{\partial x_0}{\partial t} - \int_{t_{1/2}}^{t} g_0(\bar{t},\tau)d\bar{t} + \beta\sin kt + C_1(\tau) \quad (4.3.33)$$

with

$$g_0(t,\tau) = x_0(t,\tau) + \partial C_0/\partial\tau. \quad (4.3.34)$$

In order to remove secular terms in x_1 we set the following averaging condition

$$\int_{\tau\nu}^{\tau\nu+2\pi/k} g_0(t,\tau)dt = 0 \quad (4.3.35)$$

or

$$\frac{\partial C_0}{\partial \tau} = -\frac{k}{2\pi}\int_{\tau\nu}^{\tau\nu+2\pi/k} x_0(t,\tau)dt. \quad (4.3.36)$$

Since at time $t = t_{1/2}$ the solution starts at the value $x = 2$, the initial value $C_0(0)$ is $2/3-\alpha$. The solution will leave the region A at time

$$t_m = (-\pi/2+2\pi m)/k$$

as then the line $x = 1$ is approached. This takes place as $C_0(\tau)$ reaches the value $-2/3+\alpha$. From (4.3.36) it follows that in the slow time scale this will be for

$$H(\alpha) = \frac{2\pi}{k}\int_{-2/3+\alpha}^{2/3-\alpha}\{\int_0^{2\pi/k} x_0(t;C_0)dt\}^{-1}dC_0. \quad (4.3.37)$$

It is easily verified that indeed $H(0) = 3/2-\ln 2$ and $H(2/3) = 0$. Finally, the third equation reads

$$2\frac{\partial^2 x_0}{\partial t\partial\tau} + \frac{\partial^2 x_1}{\partial t^2} + (x_0^2-1)(\frac{\partial x_1}{\partial t} + \frac{\partial x_2}{\partial t}) + 2x_0x_1(\frac{\partial x_0}{\partial \tau} + \frac{\partial x_1}{\partial t}) +$$

$$+(x_1^2+2x_0x_2)\frac{\partial x_0}{\partial t} + x_1 = 0, \quad (4.3.38)$$

so according to (4.3.34) we have

$$(x_0^2-1)x_2 = -\frac{\partial x_0}{\partial \tau} - \frac{\partial x_1}{\partial t} + \int_{t_0+\pi}^{t}\{G_0-g_1\}d\bar{t} - x_1^2x_0 \quad (4.3.39)$$

with

$$G_0 = \int_{t_0+\pi}^{t} \frac{\partial g_0}{\partial \tau}\, d\bar{t}, \quad g_1 = x_1 + \frac{\partial C_1}{\partial \tau}. \tag{4.3.40}$$

The term $\partial g_0/\partial\tau$ satisfies the averaging condition and behaves as $a + b/(x_0^2 - 1)$, therefore its integral is not secular. The averaging condition

$$\int_{\tau\nu}^{\tau\nu+2\pi} g_1(t,\tau)dt = 0 \tag{4.3.41}$$

yields a linear differential equation for C_1:

$$\frac{\partial C_1}{\partial \tau} + \frac{C_1}{2\pi}\int_{\tau\nu}^{\tau\nu+2\pi} \frac{1}{x_0^2-1}dt = -\frac{\beta}{2\pi}\int_{\tau\nu}^{\tau\nu+2\pi} \frac{\sin kt}{x_0^2-1}dt. \tag{4.3.42}$$

Substituting $t = t^{1/2} + \xi\nu^{-1/2}$ in (4.3.28) we obtain

$$C_1(0) = C_{10} = \beta + a^2(K_{1/2} + \tfrac{1}{2}), \tag{4.3.43}$$

so that

$$C_1(\tau) = \exp\{-\frac{k}{2\pi}\int_0^{\tau}\int_{\bar{\tau}\nu}^{\bar{\tau}\nu+2\pi/k} \frac{1}{x_0^2-1}dtd\bar{\tau}\}$$

$$\times[C_{10} - \frac{\beta}{2\pi}\int_0^{\tau}\exp\{\frac{k}{2\pi}\int_0^{\bar{\tau}}\int_{\tilde{\tau}\nu}^{\bar{\tau}\nu+2\pi/k} \frac{1}{x_0^2-1}dtd\tilde{\tau}\int_{\bar{\tau}\nu}^{\bar{\tau}\nu+2\pi/k} \frac{\sin kt}{x_0^2-1}dtd\bar{\tau}]$$

and

$$C_1(H(\alpha)) = q(\alpha)\{C_{10} - \beta p(\alpha)\}, \tag{4.3.44a}$$

$$p(\alpha) = \frac{k}{2\pi}\int_0^{H(\alpha)}\exp\{\frac{k}{2\pi}\int_0^{\tau}\int_{\bar{\tau}\nu}^{\bar{\tau}\nu+2\pi/k} \frac{1}{x_0^2-1}dtd\bar{\tau}\}$$

$$\times \int_{\tau\nu}^{\tau\nu+2\pi/k} \frac{\sin kt}{x_0^2-1}dtd\tau, \tag{4.3.44b}$$

$$q(\alpha) = \exp\{-\frac{k}{2\pi}\int_0^{H(\alpha)}\int_{\tau\nu}^{\tau\nu+2\pi/k} \frac{1}{x_0^2-1}dtd\tau\}. \tag{4.3.44c}$$

When the solution approaches a $\nu^{-1/2}$-neighbourhood of $(x,t) = (1,t_m)$ the expansion (4.3.28) behaves asymptotically as

$$x \approx 1 - \tfrac{1}{2}a^2(t-t_m) +$$

$$+\{-\tfrac{1}{2}a^2 + \beta - C_1(\tau) - C'_0(H)(t_m - t_{1/2} - H\nu)\}$$

$$\times\{a^2(t-t_m)\nu\}^{-1}. \tag{4.3.45}$$

From (4.3.8) we conclude that the solution for region A matches the ones from regions A_{n-m} for m and n of order $O(\nu)$ giving

$$C_1^{(m)} - I = C_1(H) + C'_0(H)(t - t_{1/2} - H\nu) \tag{4.3.46}$$

or as $I = -2\pi C'_0(H)/k$,

$$C_1^{(m)} = C_1(H) - (m - \tfrac{3}{2})I + (2\pi/k)^{-1} IH\nu. \tag{4.3.47}$$

Symmetric solutions of period $T = 2\pi(2n-1)/k$. The construction of a symmetric periodic solution of period $T = 2\pi(2n-1)/k$ with

$$x(t) = -x(t - \pi(2n-1)/k)$$

is completed as follows. From the relations (4.3.9) we derive the mapping from $K_{1/2}$ with range $(0, 2\pi I)$ to K_n within the same range: $K_n = P(K_{1/2})$. It turns

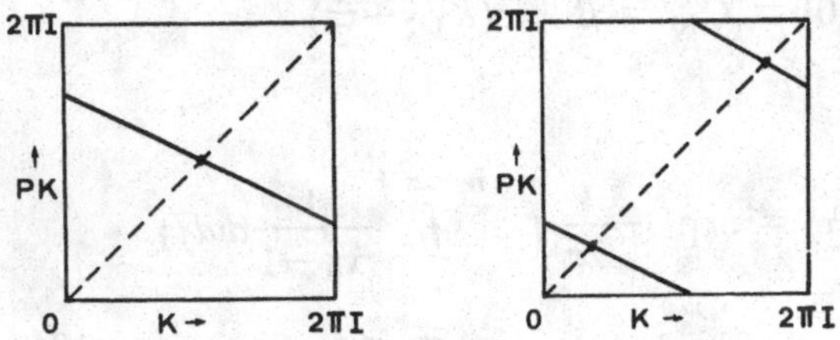

Fig.4.3.2 Discontinuous approximation of the iteration mapping

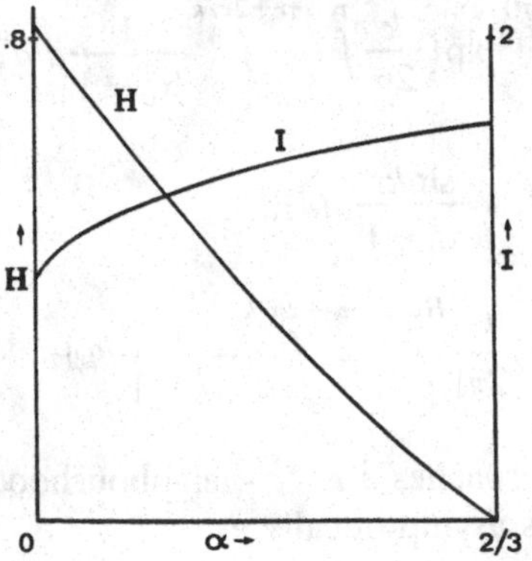

Fig.4.3.3. The period $H(\alpha)$ and characteristic function $I(\alpha)$

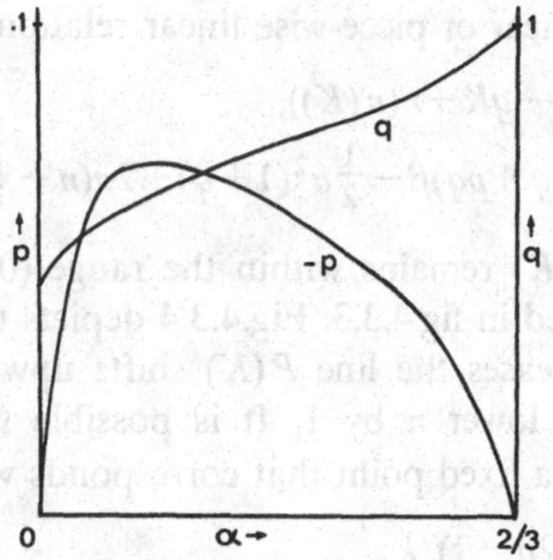

Fig.4.3.4 The auxiliary functions $p(\alpha)$ and $q(\alpha)$

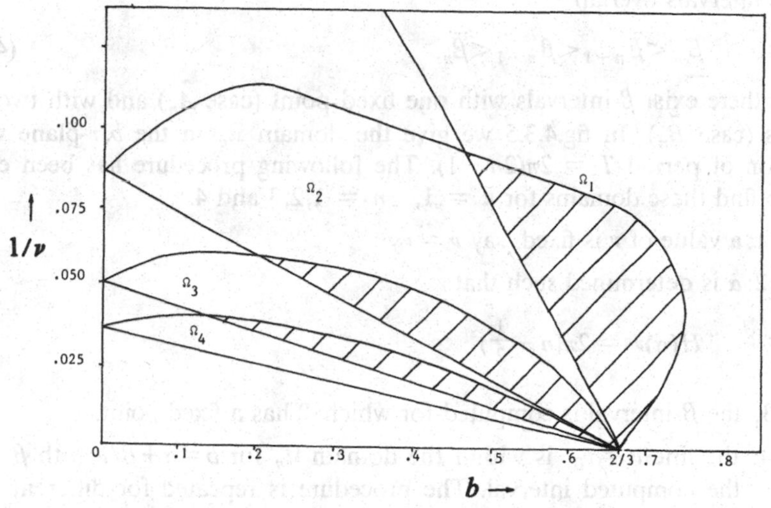

Fig.4.3.5 The domains in the parameter plane with subharmonics of period $T = 2\pi(2n-1)$

out to be a simple linear or piece-wise linear relation, see fig.4.3.2,

$$P(K) = -qK + r(n(K)), \tag{4.3.48a}$$

$$r = (1+q+pq)\beta - \frac{1}{2}a^2(1+q) + 2\pi(n-\frac{1}{2})I - kIH\nu \tag{4.3.48b}$$

with n such that $P(K)$ remains within the range $(0,2\pi I)$. The functions $I(\alpha)$ and $H(\alpha)$ are sketched in fig.4.3.3. Fig.4.3.4 depicts the auxiliary functions $p(\alpha)$ and $q(\alpha)$. As β increases the line $P(K)$ shifts upwards and as it reaches the upper boundary we lower n by 1. It is possible to compute the β-intervals $(\underline{\beta}_n, \overline{\beta}_n)$ where P has a fixed point that corresponds with a solution of period

$$T = 2\pi(2n-1)/k:$$

$$\underline{\beta}_n = \{\frac{1}{2}a^2(1+q) - h_n(\alpha,\nu)\}/s, \tag{4.3.49a}$$

$$\overline{\beta}_n = \{(\frac{1}{2}a^2 + 2\pi I)(1+q) - h_n(\alpha,\nu)\}/s, \tag{4.3.49b}$$

$$h_n = \{2\pi(n-\frac{1}{2}) - kH\nu\}I,$$

$$s = 1+q+pq.$$

These intervals overlap

$$\underline{\beta}_n < \overline{\beta}_{n+1} < \underline{\beta}_{n-1} < \overline{\beta}_n. \tag{4.3.50}$$

Thus, there exist β-intervals with one fixed point (case $\tilde{A}_n$) and with two fixed points (case $\tilde{B}_n$). In fig.4.3.5 we give the domain Ω_n in the b,ν-plane with a solution of period $T = 2\pi(2n-1)$. The following procedure has been carried out to find these domains for $k = 1$, $n = 1,2,3$ and 4.

Step 1: a value of ν is fixed, say $\nu = \nu_0$

Step 2: α is determined such that

$$H(\alpha)\nu = 2\pi(n-\frac{1}{2})$$

Step 3: the β-interval is computed for which P has a fixed point.

Step 4: the line $\nu = \nu_0$ is within the domain Ω_n for $b = \alpha + \beta/\nu$ with β within the computed interval. The procedure is repeated for different values of ν.

The domains Ω_n agree quite well with the corresponding domains of Flaherty and Hoppensteadt (1978). For $b < 1/\nu$ we used the asymptotic solution of section 4.2.1. If for $\alpha = 2/3$ we consider the range of β given by (4.3.49) we observe that this special case, studied in Grasman (1980), is completely covered by the present results. Since

$$I(2/3) = 3\sqrt{3}/\pi, \quad H(2/3) = 0,$$

and

$$p(2/3) = q(2/3) = 0,$$

the conditions on β read

$$3\sqrt{3}(\frac{11}{18}-n)<\beta<3\sqrt{3}(\frac{47}{18}-n). \tag{4.3.51}$$

For $\alpha = 0$ we have

$$I(0) = 2\pi, \quad H(0) = 3/2 - \ln 2$$

and

$$p(0) = 0, \quad q(0) = 1/2,$$

so that

$$(3-2\ln 2)\nu - 2\pi(2n-1) < 3\beta < (3-2\ln 2)\nu - 2\pi(2n-2), \tag{4.3.52}$$

which for $\beta \to \infty$ matches the conditions on this parameter given by inequalities (4.2.39) and (4.2.40).

4.3.2. Dips and slices

When $K_{1/2}$ or K_n is near one of the end points of the interval $(0, 2\pi I)$ the solution will follow a completely different path before entering a type A region. In fig.4.3.6 we explore this phenomenon by solving (4.3.1) numerically for two sets of starting values that agree upto the fifth decimal. Fig.4.3.7 gives the regions, where the solution has its own local behaviour. We consider the case where

$$K_{1/2} = \sigma \exp(-d\nu) \tag{4.3.53}$$

with $\sigma = \pm 1$. This choice of $K_{1/2}$ will produce a local behaviour near the line $x = -1$, which we indicate by dips and slices of the solution.

If for a moment we take $\sigma = 0$, then the expansion (4.3.3) of region $A_{1/2}$ remains regular and at point $t = t_{1/2}$ the solution will smoothly switch to a different expansion with a leading term

$$x_{3/2,0}(t) = 2\cos[\tfrac{1}{3}(\arccos\{\tfrac{3}{2}\sin kt + \tfrac{3}{2}\alpha - 1\} + \tfrac{4}{3}\pi)] \tag{4.3.54}$$

being the second branch of (4.3.6). This solution will hold asymptotically for the region $Z_{3/2}$ over the interval $(t_{1/2}, t_{3/2})$.

We will deal with a regular asymptotic solution $\tilde{x}(t,\nu)$ of the form (4.3.3), which has two distinct representations (4.3.4) - (4.3.10) with $C_1^{(1/2)}$ such that $K_{1/2} = 0$ for $t \leqslant t_{1/2}$. For $t \geqslant t_{1/2}$ we have a representation given by (4.3.54) and an equation for $x_{3/2,1}(t)$ of the type (4.3.7) with

$$C_1^{(3/2)} = \beta - \tfrac{1}{2}a^2. \tag{4.3.55}$$

This solution is perturbed as follows

$$x = \tilde{x}(t;\nu) + V(t;\nu). \tag{4.3.56}$$

Substitution in (4.3.1) yields for the leading terms in V:

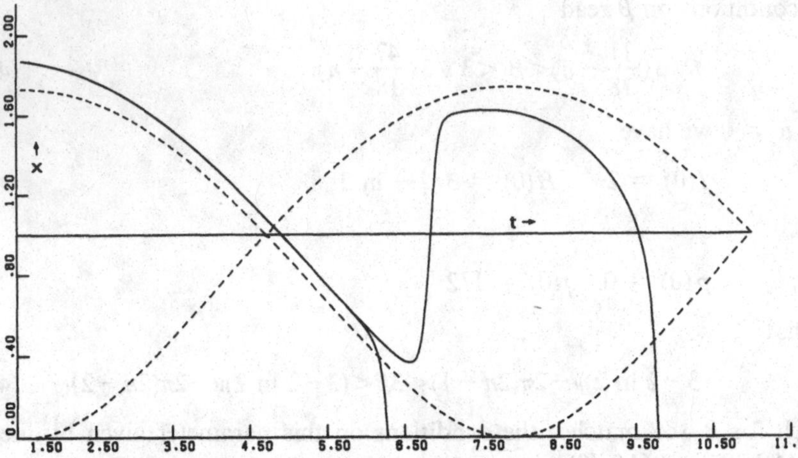

Fig.4.3.6 Two numerical solutions for $\alpha = 1/3$, $\beta = 0$, $\nu = 15$, $x'(\pi/2) = 0.0521755$: $x(\pi/2) = 1.8711914$ (dip) and $x(\pi/2) = 1.8711901$ (slice)

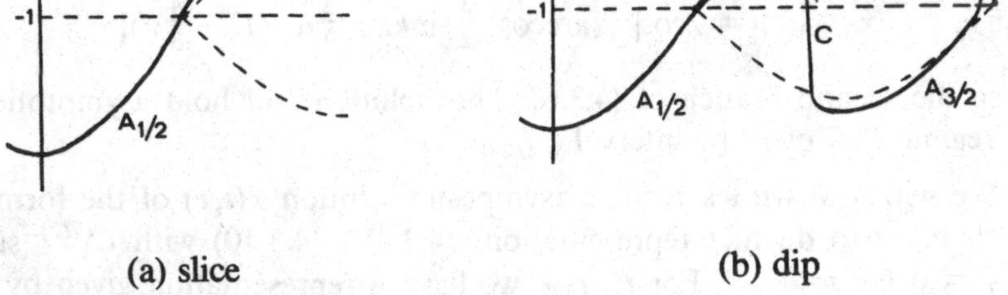

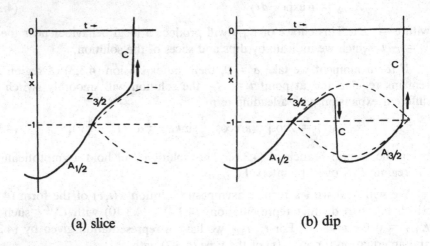

Fig.4.3.7 Regions for local asymptotic solutions in the case of dips and slices

$$\frac{d^2V}{dt^2}+\nu\frac{d}{dt}\{(x^2-1)V\} = 0. \tag{4.3.57}$$

To derive a matching relation of V with the solution for region $A_{1/2}$ given by (4.3.3) and (4.3.53), we introduce the local coordinate ξ:

$$t = t_{1/2}+\xi\nu^{-1/2}. \tag{4.3.58}$$

Using (4.3.8) we obtain

$$V\approx\frac{\sigma}{a^2\xi}\nu^{-1/2}e^{-d\nu} \quad \text{for} \quad \xi\to-\infty, \tag{4.3.59}$$

which is satisfied by

$$V = \sigma \exp\{-\nu(A(t)+d)\}[\sqrt{\pi/(2a^2\nu)}-\int_{t_{1/2}}^{t}\exp\{\nu A(\bar{t})\}d\bar{t}], \tag{4.3.60a}$$

$$A(t) = \int_{t_{1/2}}^{t}\{\tilde{x}^2(\bar{t}:0)-1\}d\bar{t}. \tag{4.3.60b}$$

This asymptotic solution breaks down as t approaches t^* satisfying

$$A(t^*) = -d. \tag{4.3.61}$$

Assuming that $t^*<t_{3/2}$ we have to introduce another boundary layer type of solution for the region C

$$x = w_0(\eta)+\nu^{-1}w_1(\eta)+\cdots, \quad \eta = (t-t^*)\nu. \tag{4.3.62}$$

Applied to Eq. (4.3.1) this yields the recurrent system

$$\frac{d^2w_0}{d\eta^2}+(w_0^2-1)\frac{dw_0}{d\eta} = 0, \tag{4.3.63}$$

$$\frac{d^2w_1}{d\eta^2}+(w_0^2-1)\frac{dw_1}{d\eta}+2w_0w_1\frac{dw_0}{d\eta} = \alpha\cos t^*,\ldots. \tag{4.3.64}$$

From (4.3.60) it follows that for $\eta\to-\infty$, w_i have to satisfy the matching conditions

$$w_0\approx\tilde{x}+\sigma\sqrt{\pi/(2a^2\nu)}\exp\{-\tilde{x}^2-1)\eta\}, \quad \tilde{x} = \tilde{x}(t^*;0), \tag{4.3.65a}$$

$$w_1\approx\frac{\alpha\cos\tilde{t}}{\tilde{x}^2-1}+\frac{1}{\tilde{x}^2-1} \tag{4.3.65b}$$

$$\times\{\frac{-\alpha\cos t^*}{(\tilde{x}^2-1}-\int_{t_n}^{t^*}\hat{x}^2(t;0)dt+\beta\sin t^*+C_{-1}^{(1/2)}\}$$

we find

$$\frac{\ln|w_0-\tilde{x}_+|}{\tilde{x}_+^2-1}+\frac{\ln|w_0-\tilde{x}_-|}{\tilde{x}_-^2-1}\frac{\ln|w_0-\tilde{x}|}{\tilde{x}^2-1} =$$

$$-\eta+\frac{\ln|-\frac{1}{2}\sigma\sqrt{\pi/(2a^2\nu)}|}{\tilde{x}^2-1}, \tag{4.3.66}$$

in which $\tilde{x}_+$ and $\tilde{x}_-$ are the two other roots of the algebraic equation

$$1/3w_0^3-w_0 = 1/3\tilde{x}^3-\tilde{x} \tag{4.3.67}$$

with $\tilde{x}_-<-1$ and $\tilde{x}_+>1$. For w_1 we have

$$\frac{dw_1}{d\nu}+(w_0^2-1)w_1 = \alpha\eta k\cos kt^* - \int_{t_{1/2}}^{t^*}\hat{x}(t;0)dt + \tag{4.3.68}$$

$$+\beta\sin kt^* - \beta + \frac{1}{2}a^2.$$

Taking $\sigma = 1$ we find that $w_0 \to \tilde{x}_+$ as $\eta\to\infty$ and matches the solution of region A, see (4.3.28), if

$$C_{10} = -\beta+\frac{1}{2}a^2-\int_{1/2}^{t^*}\hat{x}(t;0)dt. \tag{4.3.69}$$

For $\sigma = -1$ the leading term w_0 tends to the other stable root $\tilde{x}_-<-1$ and matches the asymptotic solution for region $A_{3/2}$ given by $x = \hat{x}_-(t;\nu)$ satisfying (4.3.3) with $j = 1, C_{-0}^{(3/2)} = -\alpha+3/2$ and

$$C_{-1}^{(3/2)} = C_1^{(3/2)} + \int_{t_{1/2}}^{t^*}\hat{x}(t;0)dt. \tag{4.370}$$

In the $B_{3/2}$-region (4.3.12) is valid with

$$K_{3/2} = -\int_{t_{1/2}}^{t^*}\hat{x}(t;0)dt - \int_{t^*}^{t_{3/2}}\hat{x}_-(t;0)dt. \tag{4.3.71}$$

After crossing the unstable interval $|x|<1$, the solution arrives at the region A, where it matches (4.3.28).

In Grasman (1981) the case where d is sufficiently large (including $\sigma = 0$) has been studied. Then $t^* = t_{3/2}$ and a special type of local solution is valid in the $B_{3/2}$-region. We do not present this result here as it does not affect the matching relations given above.

4.3.3 Irregular solutions

In order to analyse the mapping $P(K)$ with $K = O(\exp(-\nu))$ we shift the interval to $(-\pi I, \pi I)$ so that $K = 0$ becomes an internal point. Bringing all matching relations together, we express K_n in $K_{1/2}$ for $K_{1/2} = \pm\exp(-d\nu)$ with $d = O(1)$:

$$K_n = P(K_{1/2}) = c^{\pm}(t^*)+r(n(K_{1/2})), \quad t^* = t^*(K_{1/2}) \tag{4.3.72}$$

with

$$c^{+}(t^{*}) = q\int_{t_{1/2}}^{t^{*}} \hat{x}(t;0)dt - Ik(t^{*}-t_{1/2}), \tag{4.3.73}$$

$$c^{-}(t^{*}) = q\{\int_{t_{1/2}}^{t^{*}} \hat{x}(t;0)dt + \int_{t^{*}}^{t_{3/2}} \hat{x}_{-}(t;0)dt\} - 2\pi I, \tag{4.3.74}$$

where $t^{*} = t^{*}(K_{1/2})$ is given by (4.3.53) - (4.3.61) and $r(n(K_{1/2}))$ by (4.3.48). It is noted that this mapping $P(K)$ matches (4.3.48) for K leaving the exponentially small neighbourhood of $K = 0$, see fig.4.3.8.

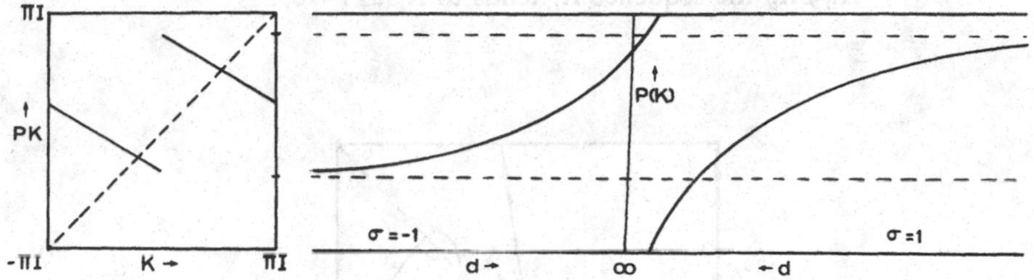

(a) discontinuous (outer) expression

(b) local (inner) expression with $K = \sigma\exp(-\nu d)$

Fig.4.3.8 The interval mapping $P(K)$

A composite asymptotic expression for the mapping P holding uniformly for $[-\pi I, \pi I]$ is constructed as follows. Let $d_{3/2} = -A(t_{3/2})$, then

$$K = -\sigma\pi I\exp\{-d(t^{*})\nu\}, \quad 0 \leqslant d \leqslant d_{3/2}, \tag{4.3.75a}$$

$$P(K) = c^{\pm}(t^{*}(K)) + r(n(K)) - qK. \tag{4.3.75b}$$

The case $\beta \in \tilde{A}_n$. In the discontinuous approximation we only found the stable fixed point; the unstable one is situated in the boundary layer, as is seen in fig.4.3.9. Note that the interval has been shifted in order to have $K = 0$ in the interior of the interval. For any starting value different from the unstable fixed point the iterated solution will approach the stable fixed point.

The case $\beta \in \tilde{B}_n$. Besides the two stable fixed points of the discontinuous approximations, there are also two unstable ones within the boundary layer. The situation is now more complicated, as it is incorrect to assume that for any starting value, not coinciding with the unstable points, the iterated solution will tend to the two stable fixed points. There exists a non-attracting subset of zero measure in which the iterated solution may go around in an irregu-

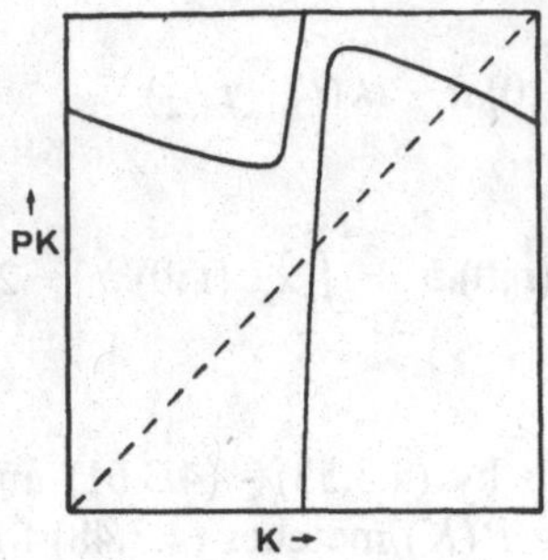

Fig.4.3.9 The interval mapping for $\beta \in \tilde{A}_n$. For any starting value $K_0 \notin K_u$ the sequence K_i tends to K_s as $i \to \infty$.

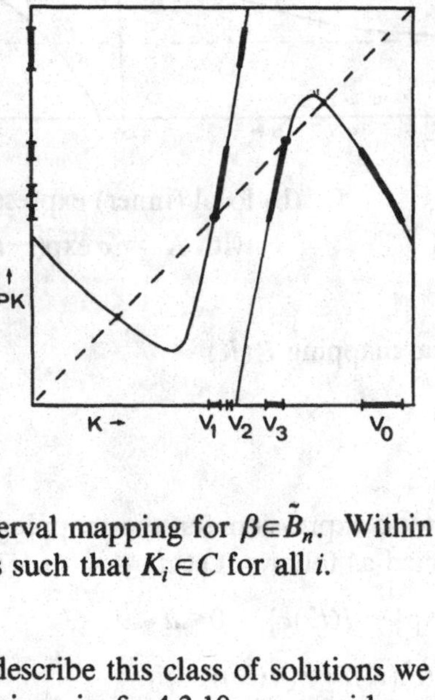

Fig.4.3.10 The interval mapping for $\beta \in \tilde{B}_n$. Within $\cup V_i$ a Cantor set C exists such that $K_i \in C$ for all i.

lar way. In order to describe this class of solutions we use symbolic dynamics, see section 4.1.2. As given in fig.4.3.10 we consider subintervals $V_i (i = 0,1,2$ and 3) and keep track of the mapping of points remaining in $\bigcup V_i$ in the transition matrix

$$M = \begin{bmatrix} 0 & 1 & 1 & 1 \\ 0 & 1 & 1 & 1 \\ 1 & 0 & 0 & 0 \\ 0 & 1 & 1 & 1 \end{bmatrix}. \tag{4.3.76}$$

If $m_{ij} = 1$ a point of V_i can be mapped in V_j, while for $m_{ij} = 0$ such a mapping is not possible. As we described in section 4.1.2, the topological subspace Σ_M consisting of all infinite sequences of the symbols 0,1,2 and 3 is introduced allowing only combinations ij for which $m_{ij} = 1$, i.e. forbidden combinations in the set of sequences are 00,10,21,22,23 and 30. It is noted that the two unstable solutions are represented as sequences of just the symbol 1 and the symbol 3, respectively. Furthermore it is seen that the interval mapping discloses the dynamics of (4.3.1) to the same extent as the annulus mapping of section 4.1.2. The iterated solutions that correspond with an element of Σ_M have zero measure. Nevertheless they give us insight in the behaviour of (4.3.1) with starting values chosen in such a way that the solution remains in $\bigcup V_i$ a large (but finite) number of iterations of P before locking into a stable subharmonic. Initially such solutions behave in the irregular way as described here and the set of starting values has a measure different from zero.

The transitional case. In the discontinuous approximation of section 4.3.1 a third type of structure remained out of sight. We are aiming at the case of one stable fixed point with the point T below the unstable fixed point S, see fig.4.3.11. Then higher order stable fixed points of the iterated mapping are possible, as pointed out by Levi (1981). His statement is based on a theorem of Newhouse and Palis (1976). In fig.4.3.11 we sketch such a solution. Next we will trace numerically one in a specific example.

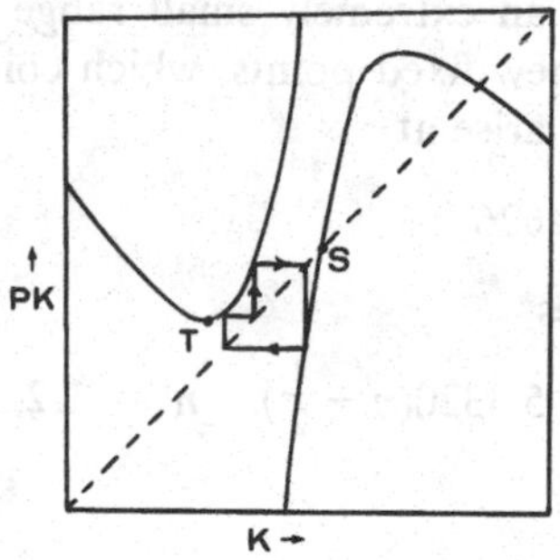

Fig.4.3.11 The interval mapping for $\beta \in \tilde{g}_n$. This case is characterized by the fact that $PK_T < PK_S$ and that P has only two fixed points. Stable irregular periodic solutions are possible.

A numerical approximation for $\alpha = 1/3$ and $\nu = 15/2$. Using analytical and numerical methods we found for $\alpha = 1/3$:

$$I = 3^{5/4}\{2E\,(\arcsin\sqrt{2/3(1-1/3\sqrt{3})},\ 1/3\sqrt{2-\sqrt{3}})$$
$$-F(\arcsin\sqrt{2/3(1-1/3\sqrt{3})},\ 1/2\sqrt{2-\sqrt{3}})$$

$$+1/3(2-\sqrt{3})\sqrt{24+14\sqrt{3}}\,\}/\pi = 1.47597, \tag{4.3.77a}$$

$$p = -0.061926, \ \ q = 0.788070, \ \ H = 0.392236, \tag{4.3.77b}$$

where

$$E(\phi,k) = \int_0^\phi \sqrt{1-k^2\sin^2\theta}\,d\theta, \tag{4.3.78a}$$

$$F(\phi,k) = \int_0^\phi (\sqrt{1-k^2\sin^2\theta})^{-1}\,d\theta. \tag{4.3.78b}$$

In order to carry out the iterations of the mapping we approximate the mapping as follows. The composite expression (4.3.75) is evaluated numerically in a set of points that have increasing density near $K = 0$. For n points we take

$$K(j) = \pm\pi I \exp\{-\frac{j\nu d_{3/2}}{n}\}. \tag{4.3.79}$$

In the computations a four point-interpolation formula is used for the points (4.3.79), where $P(K)$ is computed with a 6 decimal accuracy. Using this scheme we trace a stable fixed point of the second iterate by shifting the mapping in a vertical direction until in the iterated mapping two new fixed points arise, on of them being the stable fixed point we are looking for, see fig.4.3.12. It turns out that the stable solution has a very small domain of attraction and that it is only stable over an extremely small range of β. Therefore, we only give the value of the two new fixed points, which coincide within the accuracy we are working with. They arise at

$$K = -1.63402626 \tag{4.3.80}$$

as β takes one of the values

$$\beta = 9.3770 - 5.3320(n-\tfrac{1}{2}), \quad n = 1,2,.... \tag{4.3.81}$$

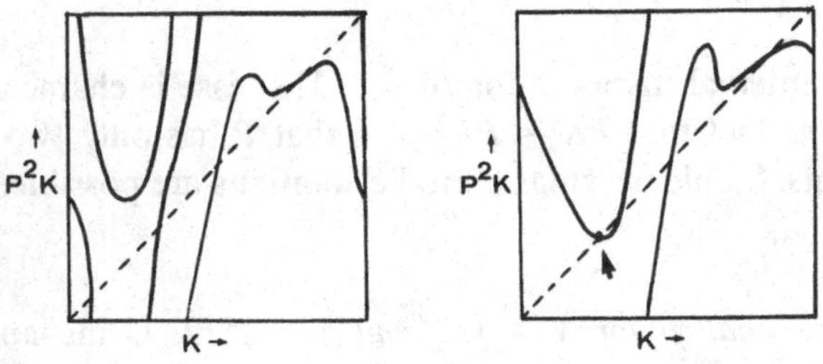

Fig.4.3.12 The occurrence of a stable irregular subharmonic

APPENDIX A: Asymptotics of some special functions

The Airy functions $Ai(z)$ and $Bi(z)$ are linearly independent solutions of the differential equation

$$\frac{d^2w}{dz^2}-zw=0, \tag{A.1}$$

see Whittaker and Watson (1927). Both functions have zero's on the negative real axis. For $Ai(z)$ the first zero occurs for $z=-\alpha=-2.33810741$. For $z>>1$ we have

$$Ai(z)=\frac{1}{2}\pi^{-1/2}z^{-1/4}e^{-\xi}\sum_{k=0}^{\infty}(-1)^k c_k\xi^{-k}, \tag{A.2a}$$

$$Bi(z)=\pi^{-1/2}z^{-1/4}e^{\xi}\sum_{k=0}^{\infty}c_k\xi^{-k} \tag{A.2b}$$

with

$$\xi=\frac{2}{3}z^{2/3},$$

$$c_0=1 \text{ and } c_k=\frac{(2k+1)(2k+3)...(6k-1)}{216^k k!},\ k=1,2,3,....$$

Moreover, for $z>>1$ the following asymptotic expansions hold

$$Ai'(z)=-\frac{1}{2}\pi^{-1/2}z^{1/4}e^{-\xi}\sum_{k=0}^{\infty}(-1)^k d_k\xi^{-k}, \tag{A.3a}$$

$$Bi'(z)=\pi^{-1/2}z^{1/4}e^{\xi}\sum_{k=0}^{\infty}d_k\xi^{-k} \tag{A.3b}$$

with

$$d_0=1 \text{ and } d_k=-\frac{6k+1}{6k-1}c_k.$$

The asymptotic behaviour at the negative real axis is given by

$$Ai(-z)=\pi^{-1/2}z^{-1/4}\{\sin(\xi+\frac{\pi}{4})\sum_{k=0}^{\infty}(-1)^k c_{2k}\xi^{-2k}+$$
$$-\cos(\xi+\frac{\pi}{4})\sum_{k=0}^{\infty}(-1)^k c_{2k+1}\xi^{-2k-1}\}, \tag{A.4a}$$

$$Bi(-z)=\pi^{-1/2}z^{-1/4}\{\cos(\xi+\frac{\pi}{4})\sum_{k=0}^{\infty}(-1)^k c_{2k}\xi^{-2k}+$$
$$+\sin(\xi+\frac{\pi}{4})\sum_{k=0}^{\infty}(-1)^k c_{2k+1}\xi^{-2k-1}\} \tag{A.4b}$$

with $z>>1$. Moreover, for $z>>1$ we have

$$Ai'(-z)=-\pi^{-1/2}z^{1/4}\{\cos(\xi+\frac{\pi}{4})\sum_{k=0}^{\infty}(-1)^k d_{2k}\xi^{-2k}+$$

$$+\sin(\xi+\frac{\pi}{4})\sum_{k=0}^{\infty}(-1)^k d_{2k+1}\xi^{-2k-1}\} \quad \text{(A.5a)}$$

and

$$Bi'(-z) = \pi^{-1/2}z^{1/4}\{\sin(\xi+\frac{\pi}{4})\sum_{k=0}^{\infty}(-1)^k d_{2k}\xi^{-2k} +$$

$$-\cos(\xi+\frac{\pi}{4})\sum_{k=0}^{\infty}(-1)^k d_{2k+1}\xi^{-2k-1}\}. \quad \text{(A.5b)}$$

Consequently, for $z << -1$ we obtain

$$-\frac{Ai'(z)}{Ai(z)} = \sqrt{-z} - \frac{1}{4z} + O(\frac{1}{-z\sqrt{-z}}). \quad \text{(A.6)}$$

For the differential equation

$$\frac{d^2w}{dz^2} + (n+\frac{1}{2}-\frac{1}{4}z^2)w = 0 \quad \text{(A.7)}$$

a standard solution exists known as the parabolic cylinder function $D_n(z)$. Other solutions are $D_n(-z)$, $D_{-n-1}(iz)$ and $D_{-n-1}(-iz)$. Any pair of these four solutions forms an independent set. The function $D_n(z)$ has no zero's on the real axis for $n \leqslant 0$. For $z >> 1$ we have

$$D_n(z) = e^{-\frac{1}{4}z^2} z^n\{1 - \frac{n(n-1)}{2z^2} + \frac{n(n-1)(n-2)(n-3)}{2.4z^4} \cdots\}. \quad \text{(A.8)}$$

At the negative real axis we have for $z << -1$

$$D_n(z) = e^{-\frac{1}{4}z^2} z^n\{1 - \frac{n(n-1)}{2z^2} + \frac{n(n-1)(n-2)(n-3)}{2.4z^2} - \cdots\} +$$

$$-\frac{\sqrt{2\pi}}{\Gamma(-n)} e^{n\pi i} e^{\frac{1}{4}z^2} z^{-n-1}\{1 + \frac{(n+1)(n+2)}{2z^2} +$$

$$+\frac{(n+1)(n+2)(n+3)(n+4)}{2.4z^4} + \cdots\}. \quad \text{(A.9)}$$

It is noted that in this formula the second expansion vanishes for n being a positive integer. From the recurrence relation

$$D_n'(z) + \frac{1}{2}zD_n(z) - nD_{n-1}(z) = 0 \quad \text{(A.10)}$$

one easily derives asymptotic expressions for $D_n'(z)/D_n(z)$ as $|z| \to \infty$.

The Gamma function $\Gamma(z)$ satisfies

$$\Gamma(z+1) = z\Gamma(z) \quad \text{(A.11)}$$

and has simple poles for $z = 0, -1, -2....$ For $z = 1,2,...$ we have

$$\Gamma(z) = (z-1)! \quad \text{(A.12)}$$

For $|z|<2$

$$\ln\Gamma(2+z) = z(1-\gamma) + \sum_{n=2}^{\infty} \frac{(-1)^n}{n}\{\zeta(n)-1\}z^n, \tag{A.13}$$

where $\zeta(s)$ denotes the Riemann zeta function

$$\zeta(s) = \sum_{k=1}^{\infty} k^{-s}. \tag{A.14}$$

For more on special functions and relaxation oscillations we refer to Zarov *et al.* (1981).

Appendix B: Asymptotic ordering and expansions

The asymptotic order of magnitude of continuous functions $\phi(x;\epsilon)$ with ϵ a small positive parameter and x restricted to an interval $I \subset R$ is measured by order functions $\delta(\epsilon)$. The following conventional symbols 'large O' and 'small o' are used to indicate the asymptotic ordering of two such functions:

$$\delta_2(\epsilon) = O(\delta_1(\epsilon)) \text{ if } \lim_{\epsilon\to 0}\delta_2/\delta_1 \text{ is bounded,}$$

$$\delta_2(\epsilon) = o(\delta_1(\epsilon)) \text{ if } \lim_{\epsilon\to 0}\delta_2/\delta_1 = 0.$$

With $\phi(x;\epsilon) = O(\delta(\epsilon))$ is meant

$$\lim_{\epsilon\to 0}\phi(x;\epsilon)/\delta(\epsilon) \text{ is bounded for all } x \in I$$

and likewise $\phi(x;\epsilon) = o(\delta(\epsilon))$ means

$$\lim_{\epsilon\to 0}\phi(x;\epsilon)/\delta(\epsilon) = 0 \text{ for all } x \in I.$$

A sequence of order functions $\delta_i(\epsilon), i = 0,1,2,...$ is called an asymptotic sequence if

$$\delta_{i+1}/\delta_i = o(1).$$

A sum of terms formed by an asymptotic sequence in the way

$$\phi^{(n)}(x;\epsilon) = \sum_{i=0}^{n} a_i(x)\delta_i(\epsilon) \tag{B.1}$$

is called an asymptotic expansion of a function $\phi(x;\epsilon)$ if

$$R^{(n)}(x;\epsilon) = \phi(x;\epsilon) - \phi^{(n)}(x;\epsilon) = o(\delta_n(\epsilon)).$$

The coefficients $a_i(x)$ are uniquely determined by the limits

$$a_0(x) = \lim_{\epsilon\to 0} \phi(x;\epsilon)/\delta_0(\epsilon),$$

$$a_i(x) = \lim_{\epsilon\to 0}\frac{\{\phi(x;\epsilon) - \phi^{(i-1)}(x;\epsilon)\}}{\delta_i(\epsilon)}, \; i = 1,2,... .$$

If one let $a_i = a_i(x;\epsilon)$, it is still required that an asymptotic sequence of order functions $\tilde{\delta}_i(\epsilon), i=1,2,...$ exists with

$$R^{(i)}(x;\epsilon) = o(\tilde{\delta}_i(\epsilon)).$$

In this way a power series expansion with respect to ϵ can be made with the coefficients still depending on $\ln\epsilon$. Moreover, the series can reordered for matching purposes. For $n \to \infty$ we have an infinite asymptotic expansion, which does not necessarily form a convergent series.

The above asymptotic ordering and expansions also aply to the argument of a function $\phi(x)$ for $x \to 0$ as well as to the case that the parameter or argument is large. More on asymptotic ordering can be found in Eckhaus (1979) and Kevorkian and Cole(1981).

Appendix C: Concepts of the theory of dynamical systems

A state of a physical system is information about this system at a given time contained in a number of variables. It is supposed that the variables are continuous in time and that only a finite number of them are taken into consideration:

$$x(t) = \{x_1(t), x_2(t), ..., x_n(t)\}.$$

The space of states $\mathbf{R}^n$ is called the *phase space.* We also use the notion "*state space*" in order to avoid confusion at instances when we also speak of the space of phases of a system of oscillators.

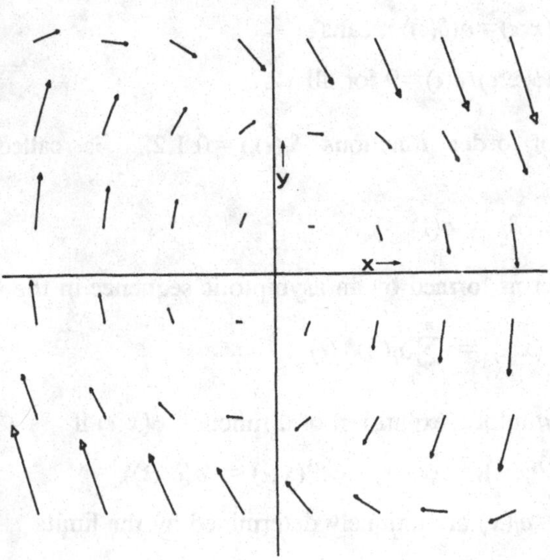

Fig.C1 The vector field of the Van der Pol oscillator (2.1.2) with $\nu=1$

When the change in time of variables can be expressed in the variables themselves, we obtain an autonomous system of differential equations

$$\frac{dx}{dt} = f(x). \tag{C.1}$$

The continuous mapping

$$f:\mathbf{R}^n \rightarrow \mathbf{R}^n \tag{C.2}$$

is called the *vector field:* to each point x a vector $f(x)$ is assigned. A diagram of vectors on a grid in phase space visualizes the global direction of change, see fig.C1.

Let at time $t=0$ a system be in a given state x_0. Then the states for $t<0$ and $t>0$ are uniquely determined. Such a solution of (C.1) is called a *trajectory:*

$$x(t) = \phi_t(x_0). \tag{C.3}$$

The collection of mappings

$$\phi_t\colon \mathbf{R}^n \rightarrow \mathbf{R}^n,\ t \in \mathbf{R}$$

is called the *dynamical system* or the *flow.*

A set $I \subset \mathbf{R}^n$ is *invariant,* if for every $x \in I, \phi_t(x)$ is in I for $t \in \mathbf{R}$.

A simply connected set $M \subset \mathbf{R}^n$ is called a *manifold* of dimension $m (\leqslant n)$, if for any $x \in M$ the set of points U on M, in a neighbourhood of x, is homeomorphic to a neighbourhood of a point in $\mathbf{R}^m$. Moreover, M can be covered by a countable number of sets U.

In many cases a $(n-p)$-dimensional manifold is determined by a set of p algebraic equations of the type

$$g_i(x) = 0, \quad i = 1,\ldots,p. \tag{C.4}$$

An invariant I is another example of a manifold in the phase space $\mathbf{R}_n$.

A *transversal intersection* Σ of a m-dimensional invariant I is a $(m-1)$-dimensional manifold contained in I, such that for any $x \in \Sigma$ a trajectory $\phi_t(x)$ exists with $\phi_t(x) \notin \Sigma$ for $0<|t|<\delta$, where δ is a sufficiently small positive number.

Let us assume that for every $x \in \Sigma$ a $t>0$ exists such that $\phi_t(x) \in \Sigma$. Then we may define the Poincaré mapping $P:\Sigma \rightarrow \Sigma$ by

$$P\colon x \mapsto \phi_t(x) \tag{C.5}$$

with $\phi_\tau(x) \notin \Sigma$ for all $0<\tau<t$. It is noted that P is a continuous mapping. It is called a *discrete dynamical system* or an *iteration map.*

A point $\bar{x}$ is a *limit point* of a trajectory $x(t)=\phi_t(x)$, if a sequence t_n exists with $t_n \rightarrow \infty$ as $n \rightarrow \infty$ such that $x(t_n) \rightarrow \bar{x}$ as $n \rightarrow \infty$.

A dynamical system, with trajectories that remain in a bounded domain of the phase space, may have the following sets of limit points:

a. *an equilibrium* $\bar{x}$ satisfying $f(\bar{x})=0$,
b. *a limit cycle* being a closed trajectory homeomorphic to the circle S^1,
c. a manifold homeomorphic to the *torus* $T^m = S^1 \times ... \times S^1$, $m<n$,
d. *a strange attractor* being a set of points of fractional dimension. The order, inwhich a neighbourhood of each of these points is visited, is irregular and varies strongly with x_0.

In fig. C2 we sketch a limit set $\omega \subset \mathbb{R}^3$ being a torus T^2.

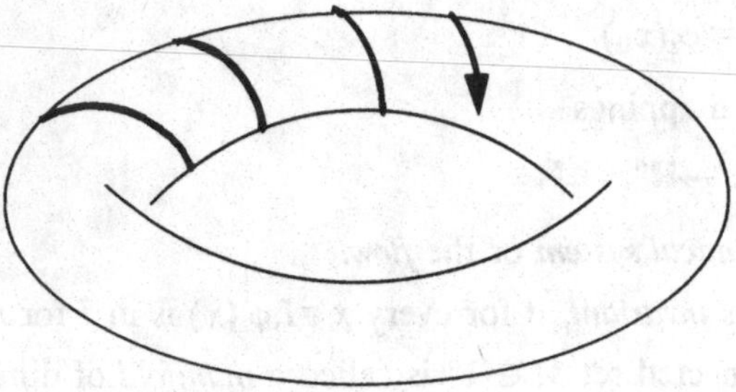

(a) torus in phase space

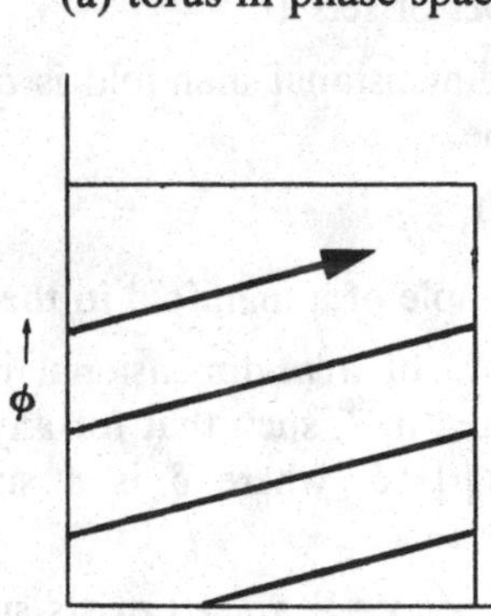

(b) trajectory on torus in angular variables

Fig.C2 A torus as a limit set ω. For starting values near ω trajectories tend to ω. A trajectory that starts in ω remains in ω and fills the torus without closing its path, because the angular velocities have an irrational ratio.

A vector field is *dissipative* if $\nabla.f(x)<0$. It has a limit set ω with dimension smaller than n.

The Poincaré mapping $P{:}\Sigma\rightarrow\Sigma$ (or P^k) may have a stable fixed point which corresponds to a limit cycle of the continuous dynamical system. A torus T^2 produces a closed curve in Σ and a strange attractor a Cantor set. For each point of a *Cantor set* a sufficiently small neighbourhood U exists such that no other point of the set is contained in U.

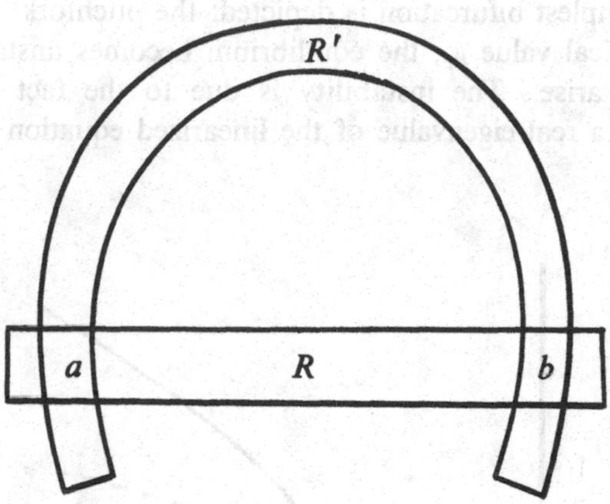

Fig.C3 The horse-shoe map. The rectangle R is mapped into the "horse-shoe" R'. The intersection a is stretched all over the horse-shoe: its image has points in a as well as in b. The intersection b is stretched likewise by the mapping. A point $x_0 \in a \cup b$ exists such that $P^k x_0$ is in a or b for $k=1,2,...$ in a given arbitrary order.

Let a dissipative dynamical system with a 3-dimensional invariant I have a 2-dimensional transversal intersection Σ such that the strip $ABCD$ is mapped into $A'B'C'D'$ by the Poincaré mapping defined above, see fig. C3. This is the *horse-shoe map* of Smale (1967). It is noted that because of the dissipation the surface of the image is smaller than that of the original strip. Thus the compression in the direction AB is larger than the stretching in the direction AD. The original and the image has two areas in common: a and b. By the mapping the area a is deformed in such a way that its image again has points in a as well as in b. The same holds for b. A repeated mapping of a segment of the image with lies in a or b yields an image that points in a as well as in b. Given an arbitrary sequence of a's and b's, say *abbbaaba...*, we may find a starting value in a or b such that repeated mappings yields points in a or b in the same prescribed order. Thus, the invariant I contains a limit set ω with a random nature as is seen from the intersections with the transversal Σ. For more information about such limit sets we refer to Guckenheimer and Holmes (1984). Henon (1976) gave the following simple discrete dynamical system in $\mathbb{R}^2$, that has a strange attractor

$$P:(x_1,x_2)\mapsto(x_2+1-1.4x_1^2, 0.3x_1). \tag{C.6}$$

Let us consider the case where a dynamical system depends on a parameter μ such that at certain values of μ the number of limit set ω changes. This phenomenon is known as *bifurcation* of solutions, see Iooss and Joseph (1980). In fig. C4 the simplest bifurcation is depicted: the pitchfork. As the parameter μ passes the critical value μ_c, the equilibrium becomes unstable and two new stable solutions arise. The instability is due to the fact that at a critical parameter value a real eigenvalue of the linearized equation becomes positive.

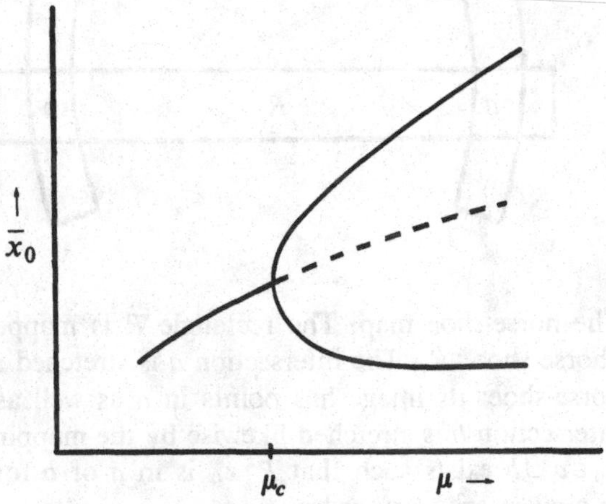

Fig.C4 Pitchfork bifurcation. Stationary solutions of $dx/dt=f_u(x)$ are presented as a function of μ. At $\mu=\mu_c$, the bifurcation point, the solution $\bar{x}_0(\mu)$ becomes unstable and two new stable solutions branch off.

If two complex conjugate eigenvalues cross the imaginary axis a periodic solution arises. This type of bifurcation is called a *Hopf bifurcation.* A discrete dynamical system or a Poincaré mapping at a transversal intersection may also have bifurcating solutions. A fixed point of a Poincaré mapping corresponds with a periodic solution of the continuous dynamical system. After bifiurcation a fixed point of P^2 may branch off, see fig. C5. The system then doubles its period. Repeated *period doubling* will lead to chaotic solutions. In section 2.6 we discuss the logistic map which has this property.

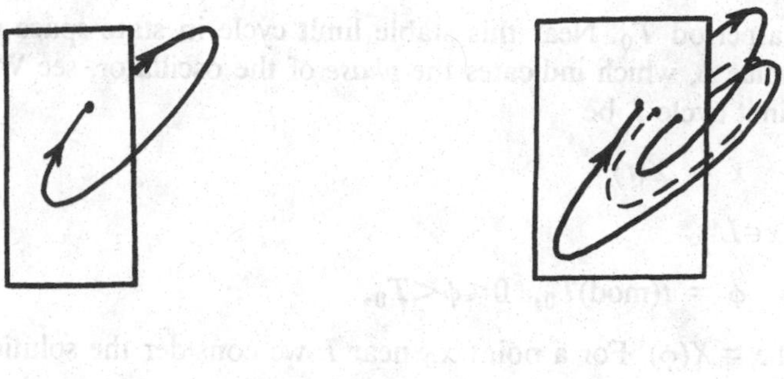

(a) $\mu<\mu_c$ (b) $\mu>\mu_c$

Fig.C5 Period doubling bifurcation. The fixed point of the Poincaré mapping becomes unstable for $\mu=\mu_c$ and two stable fixed points of P^2 arise.

When the change of variables of a dynamical system also depends on time t, the system of differential equations is nonautonomous

$$\frac{dx}{dt} = f(x,t). \tag{C.7}$$

We are in particular interested in the case that the right-hand side is periodic in t with period T. Let $x=x(t)$ be a solution with $x(\theta)=x_0, 0\leqslant\theta<T$. Then we may consider the mapping

$$P_\theta : x_0 \mapsto x(\theta+T). \tag{C.8}$$

The discrete dynamical system $P_\theta:\mathbb{R}^n\to\mathbb{R}^n$ is equivalent with the continuous system (C7).

Let us now consider the problem of a periodically forced oscillator of the type

$$\frac{dx}{dt} = f(x,t;\delta), \tag{C.9}$$

where δ is a measure for the intensity of the forcing.

The unperturbed oscillator satisfies

$$\frac{dx}{dt} = h(x),\ \ h(x) = f(x,t;0). \tag{C.10}$$

and has a period T_0. Near this stable limit cycle in state space points can be given a value ϕ, which indicates the *phase* of the oscillator, see Winfree (1980). Let the limit cycle L be

$$x = X(t),$$

then for $x \in L$

$$\phi = t(\text{mod})T_0, \quad 0 \leqslant \phi < T_0, \tag{C.11}$$

such that $x = X(\phi)$. For a point x_0 near L we consider the solution $x(t)$ of the initial value problem

$$\frac{dx}{dt} = h(x), \quad x(0) = x_0, \tag{C.12}$$

and find ϕ such that $x(t) \rightarrow X(t+\phi)$ for $t \rightarrow \infty$.

For δ sufficiently small, trajectories of the perturbed system remain near L and we may as well consider a system of differential equations defined on a torus:

$$\frac{d\phi}{dt} = 1 + \delta g(\theta), \quad \phi(0) = \phi_0, \tag{C.13a}$$

$$\frac{d\theta}{dt} = 1, \quad \theta(0) = 0. \tag{C.13b}$$

The *rotation number* is defined as

$$\rho = \lim_{k \rightarrow \infty} \frac{\phi(kT)}{kT_0}. \tag{C.14}$$

It is independent of ϕ_0. If for two integers p and q the quantity pT is sufficiently close to qT_0, a periodic solution may exist depending on $g(\theta)$. The rotation number of this solution is q/p. Its period equals pT: the oscillator gets *entrained* by the periodic forcing term.

Appendix D: Stochastic differential equations and diffusion approximations

A *stochastic process* is a system of which the state vector is a time dependent random variable $X(t)$. Let for $t > t_0$ the probability distribution $p(x,t)$ of the state vector follow from the distribution at t_0 and be independent of the history for $t < t_0$, then the stochastic process is called a *Markov process.* It satisfies the Chapman-Kolmogorov equation

$$p(x^{(2)},t_2|x^{(0)},t_0) = \int p(x^{(2)},t_2|x^{(1)},t_1)\, p(x^{(1)},t_1|x^{(0)},t_0)\, dx^{(1)}, \tag{D.1}$$

where $p(x^{(j)},t_j|x_i^{(i)}t_i)$ denotes the conditional probability of the system of being in state $x^{(j)}$ at time t_j given it is in $x^{(i)}$ at time t_i before. A Markov process is continuous in time if for any $\epsilon > 0$

$$\lim_{\Delta t \to 0} \frac{1}{\Delta t} \int_{|x-y|>\epsilon} p(x,t+\Delta t|y,t)dx = 0 \tag{D.2}$$

uniformly in y,t and Δt. A Markov process is stationary if for any τ the processes $X(t)$ and $X(t+\tau)$ have the same statistics. That is the conditional probability depends on the time difference:

$$p(x^{(j)},t_j|x^{(i)},t_j)=p(x^{(i)},t_j-t_i|x^{(i)},0).$$

Let the vector function $x(t)$ with $t \in \mathbf{R}$ be a realization of a stationary continuous Markov process $X(t)$. The autocorrelation matrix $G(t)$ is defined by

$$G_{ij}(t) = \int_{-\infty}^{\infty} x_i(t)x_j(0)dt. \tag{D.3}$$

The spectral matrix $S(\omega)$ is its Fourier transform:

$$S_{ij}(\omega) = \frac{1}{2\pi} \int_{-\infty}^{\infty} e^{-i\omega t} G_{ij}(t)dt. \tag{D4}$$

If for a 1-dimensional process $\xi(t)$ the autocorrelation function is

$$G(t)=\delta(t) \text{ and } E\{\xi(t)\}=0$$

with $\delta(t)$ the Dirac delta function, then $\xi(t)$ is called a white noise process: $S(\omega)=1/(2\pi)$, so all frequenties have equal intensity. The time-integrated white noise process is called the Wiener process

$$W(t) = \int_0^t \xi(s)ds \tag{D.5}$$

with the Wiener increment satisfying

$$dW(t) = W(t+dt) - W(t) = \xi(t)dt. \tag{D.6}$$

A dynamical system perturbed by n white noise processes takes the form of a stochastic vector differential equation

$$dX_i = f_i(X)dt + \sum_{j=1}^{n} \sigma_{ij}(X)dW_j, \quad i=1, \cdots ,n, \tag{D.7}$$

where $\sigma(X)$ denotes the diffusion matrix. Using the Chapman-Kolmogorov equation (D.1) one derives a partial differential equation, known as the Fokker-Planck equation or the forward Kolmogorov equation,

$$\frac{\partial p}{\partial t} = \frac{1}{2} \sum_{i,j=1}^{n} \frac{\partial^2}{\partial x_i \partial x_j} \{a_i(x)p\} - \sum_{i=1}^{n} \frac{\partial}{\partial x_i} \{f_i(x)p\}, \tag{D.8}$$

or

$$\frac{\partial p}{\partial t} = M_x p,$$

with

$$a_{ij} = \sum_{k=1}^{n} \sigma_{ik}\bar{\sigma}_{kj}.$$

The function $p(x,t|y,t')$ is the conditional probability of a *diffusion process* approximating the Markov process $X(t)$. Let at a time $t=0$ the probability function have an initial value $p_0(x)$. Then we only need to impose the proper boundary conditions for p in order to make the diffusion problem (D.8) well-posed. We assume that from the boundary $\partial\Omega$ of a domain Ω in state space the system cannot return in Ω. For such an absorbing boundary the condition is

$$p(x,t|y,t')=0 \quad \text{for } x\in\partial\Omega.$$

The conditional probability of the diffusion process also satisfies a partial differential equation in y, the backward Kolomogorov equation

$$\frac{\partial p}{\partial t}=\frac{1}{2}\sum_{i,j=1}^{n} a_{ij}(y)\frac{\partial^2 p}{\partial y_i \partial y_j}+\sum_{i=1}^{n} f_i(y)\frac{\partial p}{\partial y_i} \tag{D.9}$$

or

$$\frac{\partial p}{\partial t}=L_y p,$$

where the elliptic operator L_y is the formal adjoint of M_y, see (D.8).

Let $X(0)=y$ and $g(x,y,t)$ the probability of exit from Ω near x before time t. Then $g(x,y,t)$ satisfies the diffusion equation

$$\frac{\partial g}{\partial t}=L_y g \tag{D.10}$$

with initial values

$$g(x,y,0)=0 \quad \text{for } y\in\Omega \tag{D.11}$$

and boundary condition

$$g(x,y,t)=\delta_s(x-y) \quad \text{for} \quad x,y\in\delta\Omega, \tag{D.12}$$

where $\delta_s(x-y)$ is a delta function defined on $\partial\Omega$. Letting $t\to\infty$ we obtain the probability of exit through x satisfying the stationary equation

$$L_y g=0 \tag{D.13}$$

with boundary condition (D.12).

A system, that initially is in y, is still in Ω at time $t>0$ with a probability

$$G(y,t)=\int_{\Omega} p(x,t|y,0)dx.$$

satisfying

$$\frac{\partial G}{\partial t} = L_y G \tag{D.14}$$

with initial values and boundary conditions

$$G(y,0)=1 \quad \text{for} \quad y \in \Omega,$$

$$G(y,t)=0 \quad \text{for} \quad y \in \partial\Omega.$$

Let the stochastic variable T be the time of exit, then

$$G(y,t) = \text{Prob } (T \geqslant t).$$

The expected value of any function of T is

$$E\{f(T)\} = -\int_0^\infty f(t) dG(y,t). \tag{D.15}$$

Consequently, the statistical moments $T_n(y)$ of T are found from (D.15) by taking $f(t)=t^n$, so using partial integration we obtain

$$T_n(y) = \int_0^\infty t^{n-1} G(y,t) dt.$$

The first moment, being the mean exit time, satisfies

$$L_y T_1 = -1 \quad \text{in } \Omega,$$

$$T_1(y) = 0 \quad \text{at } \partial\Omega.$$

The higher moments form a recurrent system:

$$L_y T_n = -nT_{n-1} \text{ in } \Omega,$$

$$T_n = 0 \text{ at } \partial\Omega.$$

LITERATURE

ANDERSEN, C.M., and J. GEER (1982). *Power series expansions for the frequency and period of the limit cycle of the Van der Pol equation,* SIAM J. Appl. Math. **42**, p.678-693.

ANDERSON, P.M. and A.A. FOUAD (1977). *Power System Control and Stability,* Iowa State Univ. Press, Ames.

ALLEN, T. (1983). *On the arithmetic of phase locking: coupled neurons as a lattice on R^2,* Physica **6D**, p.305-320.

ALSEDÁ, IL., J. LLIBRE and R. SERRA (1983). *Bifurcations for a circle map family associated with the Van der Pol equation,* Preprint Universitat Autònoma, Barcelona.

ANAND , K.K. (1983). *On relaxation oscillations governed by a second order differential equation for a large parameter and with a piecewise linear function,* Canad. Math. Bull. **26**, p.80-91.

ARGÉMI, J. and B. ROSSETTO (1983). *Solutions periodiques discontinues pour l'approximation singulière d'un modèle neurophysiologique dans R^4 - une metaphore dans R^3 avec chaos,* J. Math. Biol. **17**, p.67-92.

AUCHMUTY, J.F.G. and G. NICOLIS (1976). *Bifurcation analysis of reaction diffusion equations - III. Chemical oscillations,* Bull. Math. Biol. **38**, p.325-350.

BAVINCK, H., and J. GRASMAN (1974). *The method of matched asymptotic expansions for the periodic solution of the Van der Pol equation,* Int. J. Nonlin. Mech. **9**, p.421-434.

BÉLAIR, J. (1983). *Phase Locking in Linearly Coupled Relaxation Oscillators.* Ph. D. Thesis, Cornell Univ., Univ. Microfilms Int., Ann Arbor.

BÉLAIR, J. and P. HOLMES (1984). *On linearly coupled relaxation oscillations,* Quart. Appl. Math. **42**, p.193-219.

BLEVINS, R.D. (1977). *Flow-induced Vibration.* Van Nostrand, New York.

BLOM, J.G., R. de BRUIN, J. GRASMAN and J.G. VERWER (1981). *Forced prey-predator oscillations,* J. Math. Biol. **12**, p.141-152.

BOGOLIUBOV, N.N., and I.A. MITROPOLSKY (1961). *Asymptotic Methods in the Theory of Nonlinear Oscillations,* Gordon and Breach, New York.

BOJADZIEV, G. (1984). *Population modelling by second order differential systems,* in Mathematical Modelling in Science and Technology, The Fourth Int. Confer., Zürich, Aug. 1983, Pergamon Press, New York.

BONHOEFFER, K.F. (1948). *Activation of passive iron as a model for the excitation of nerve,* J. Gen. Physiol. **32**, p.69-91.

BONNILLA L.L., and A. LIÑÁN (1984). *Relaxation oscillations, pulses, and travelling waves in the diffusive Vollterra delay differential equation,* SIAM J. Appl. Math. **44**, p.369-391.

BRANDSTATTER, J.J. (1963). *An Introduction to Waves, Rays and Radiation in Plasma Media,* McGraw-Hill, New York.

CALLOT, J-L., F. DIENER et M. DIENER (1978). *Le problème de la "chasse au canard",* C.R. Acad. Sc. Paris, Sér. A **286**, p.1059-1061.

CARRIER, G.F., and J.A. LEWIS (1953). *The relaxation oscillations of the Van der Pol oscillator,* Adv. in Appl. Mech., Acad. Press, New York **3**, p.12-16.

CARTWRIGHT, M.L. (1950). *Forced oscillation in nonlinear systems,* in Contrib. to the Theory of Nonlinear Oscillations, S. Lefschetz (ed.), Annals of Math. Studies **20**, p.149-242.

CARTWRIGHT, M.L. (1952). *Van der Pol's equation for relaxation oscillation,* in Contrib. to the Theory of Nonlinear Oscillations, Vol.II, S. Lefschetz (ed.), Annals of Math. Studies **29**, p.3-18.

CARTWRIGHT, M.L., and J.E. LITTLEWOOD (1947). *On nonlinear differential equations of the second order II,* Ann. of Math. **48**, p.472-494.

CARTWRIGHT, M.L., and J.E. LITTLEWOOD (1951). *Some fixed point theorems,* Ann. of Math. **54**, p.1-37.

CHANCE, B., E.K. GHOSH and B. HESS, eds., (1973). *Biological and Biochemical Oscillators,* Acad. Press, New York.

CHOW,. P.-L., and W.C. TAM (1976). *Periodic and traveling wave solutions to Volterra-Lotka equation with diffusion,* Bull. Math. Biol. **38**, p.643-658.

CLENSHAW, C.W. (1966). *The solution of Van der Pol's equation in Chebyshev series,* in Numerical Solutions of Nonlinear Differential Equations, D. Greenspan (ed.), Wiley, New York, p.55-63.

CODDINGTON, E.A., and N. LEVINSON (1955). *Theory of Ordinary Differential Equations,* McGraw-Hill, New York.

COHEN, A.H., P.J. HOLMES and R.H. RAND (1982). *The nature of the coupling between segmental oscillators of the lamprey spinal generator for locomotion: a mathematical model,* J. Math. Biol. **13**, p.345-369.

COLE, J.D. (1968). *Perturbation Methods in Applied Mathematics,* Blaisdell, Waltham, Mass.

COLLET, P. and J.-P. ECKMANN (1980). *Iterated Maps on the Interval as Dynamical Systems,* Birkhäuser, Basel.

COPSON, E.T. (1948). *An Introduction to the Theory of Functions of a Complex Variable,* Oxford Univ. Press, London.

CRONIN, J. (1977). *Some mathematics of biological oscillations,* SIAM Rev, **19**, p.100-138.

CRUTCHFIELD, J.P., B.A. HUBERMAN, N.H. PACKARD (1980). *Noise phenomena in Josephson junctions,* Appl. Phys. Lett. **37**, p.750-751.

DADFAR, M.B., J. GEER and C.M. ANDERSEN (1984). *Perturbation analysis of the limit cycle of the free Van der Pol equation,* SIAM J. Appl. Math. **44**, p.881-895.

DEPRIT, A. and A.R.M. ROM (1967). *Asymptotic representation of Van der Pol's equation for small damping coefficients,* Z. Angew. Math. Phys. **18**, p.736-747.

DIENER, F. (1980). *Les canards de l' equation* $\ddot{y}+(\dot{y}+a)^2+y=0$, Report Ser. Math. Pures et Appl., I.R.M.A. Strasbourg.

DMITRIEV, A.S. (1983). *Chaos in a driven self-oscillating system,* Radiophysica **24**, p.1081-1086 (Russian).

DORODNICYN, A.A. (1962). *Asymptotic solution of the Van der Pol equation,* Am. Math. Soc. Transl. Series 1, **4**, p.1-23.

DUTT, R. (1976). *Application of Hamilton-Jacobi theory to the Lotka-Volterra oscillator,* Bull. Math. Biol. **38**, p.459-465.

EBELING, W., HERZEL and E.E. SEL'KOV (1985). *Theory of stochastic biochemical oscillations,* in Proc. 16th FEBS Congress, VNU Science Press, 443-449.

EBELING,W., H. Herzel, W. RICHERT and L. SCHIMANSKY-GEIER (1986). *Influence of noise on Duffing-Van der Pol oscillators,* Z. Angew. Math. Mech. **66**, p. 141-146.

ECKHAUS, W. (1979). *Asymptotic Analysis of Singular Perturbations,* North Holland Publ., Amsterdam.

EL-ABBASY, E.M., E.M. JAMES (1983). *Stable subharmonics of the forced Van der Pol equation,* IMA J. Appl. Math. **31**, p.269-279.

ENGEL-HERBERT, H.,W. EBELING and H.HERZEL (1985). *The influence of fluctuations on sustained oscillations,* in RENSING and JAEGER, eds., (1985), p.144-152.

ECKHAUS, W. (1983). *Relaxation oscillations including a standard chase on French ducks,* in Asymptotic Analysis II, F. Verhulst (ed.), Springer Lecture Notes in Math. **985**, p.449-494.

ERMENTROUT, C.B. (1981). *n:m phase-locking of weakly coupled oscill-tors,* J. Math. Biol. **12**, p.327-342.

ERMENTROUT, G.B. (1985). *Synchronization in a pool of mutually coupled oscillators with random frequencies,* J. Math. Biol. **22**, p.1-10.

FIFE, P.C. (1977). *Mathematical aspects of reacting and diffusing systems,* Springer Lecture Notes in Biomath. **28**.

FITZHUGH, R. (1955). *Mathematical models of threshold phenomena in the nerve membrane,* Bull. of Math. Biophys. **17**, p.257-278.

FLAHERTY, J.E., and F.C. HOPPENSTEADT (1978). *Frequency entrainment of a forced Van der Pol oscillator,* Stud. Appl. Math. **18**, p.5-115.

FLANDERS, D.A., and J.J. STOKER (1946). *The limit case of relaxation oscillations,* in Studies in Nonlinear vibration Theory, Inst. for Math. and Mech., New York.

FRANCK, U.F. (1985). *Spontaneous temporal and spatial phenomena in physicochemical systems,* in RENSING and JAEGER, eds., (1985). p.2-12.

GARDINER, C.W. (1983). *Handbook of Stochastic Methods for Physics Chemistry and the Natural Sciences,* Springer Series in Synergetics **13**.

GLASS, L., and M.C. MACKEY (1979). *A simple model for phase locking of biological oscillators,* J. Math. Biol. **7**, p.339-352.

GLASS, L., and R. PEREZ (1982). *Fine structure of phase locking,* Phys. Rev. Letters **48**, p.1172-1775.

GOLDBETER, A. (1980). *Models for oscillations and excitability in biochemical systems,* in Mathematical models in molecular and cellular Biology, L.A. Segel (ed.), Cambridge Univ. Press, p.248-291.

GOLLUB, J.P., T.O. BRUNNER and B.G. DANLY (1978). *Periodicity and chaos in coupled nonlinear oscillators,* Science **200**, p.48-50.

GOLLUB, J.P., E.J. ROMER and J.E. SOCOLAR (1980). *Trajectory divergence for coupled relaxation oscillations: measurements and models,* J.Stat. Phys. **23**, p. 321-333.

GRASMAN, J. (1980a). *Relaxation oscillations of a Van der Pol equation with large critical forcing term,* Quart. Appl. Math. **38**, p.9-16.

GRASMAN, J. (1980b). *On the Van der Pol relaxation oscillator with a sinusoidal forcing term,* Math. Centre, Amsterdam, Report TW 207.

GRASMAN, J. (1981). *Dips and slidings of the forced Van der Pol relaxation oscillator,* Math. Centre, Amsterdam, Report TW 214.

GRASMAN, J. (1984). *The mathematical modeling of entrained biological oscillators,* Bull. Math. Biol. **46**, p.407-422.

GRASMAN, J., and M.J.W. JANSEN (1979). *Mutually synchronized relaxation oscillators as prototypes of oscillating systems in biology,* J. Math. Biol. **7**, p.171-197.

GRASMAN, J., M.J.W. JANSEN and E.J.M. VELING (1976). *Asymptotic methods for relaxation oscillations,* in Proceedings of the third Scheveningen Conference on Differential Equations, W. Eckhaus and E.M. de Jager (eds.), North-Holland Math. Studies **31**, p.93-111.

GRASMAN, J., H. NIJMEIJER and E.J.M. VELING (1984). *Singular perturbations and a mapping on an interval for the forced Van der Pol relaxation oscillator,* Physica **13D**, p.195-210.

GRASMAN, J., and J.B.M. ROERDINK (1986). *Stochastic and chaotic relaxation oscillations,* Report Centre for Math. and Computer Science,

Amsterdam.

GRASMAN, J., and E. VELING (1973). *An asymptotic formula for the period of a Volterra-Lotka system,* Math. Biosci **18**, p.185-189.

GRASMAN, J., and E.J.M. VELING (1979). *Asymptotic methods for the Volterra-Lotka equations,* in Asymptotic Analysis, F. Verhulst (ed.), Springer Lecture Notes in Math. **711**, p.146-157.

GRASMAN, J., E.J.M. VELING and G.M. WILLEMS (1976). *Relaxation oscillations governed by a Van der Pol equation with periodic forcing term,* SIAM J. Appl. Math. **31**, p.667-676.

GUCKENHEIMER, J., (1980). *Symbolic dynamics and relaxation oscillations,* Physica **1D**, p.227-235.

GUCKENHEIMER, J., and P. HOLMES (1984). *Nonlinear Oscillations, Dynamical Systems and Bifurcation of Vector Fields,* Appl. Math. Sci. **42**, Springer-Verlag, Berlin.

GUEVARA, M.R. and L. GLASS (1982). *Phase locking, period doubling bifurcation and chaos in a mathematical model of a periodically driven oscillator: a theory for the entrainment of biological oscillators and the generation of cardiac dysrhythmias,* J. Math. Biol. **14**, p.1-24.

GUEVARA, M.R., L. GLASS and A. SHRIER (1981). *Phase locking, period-doubling bifurcations, and irregular dynamics in periodically stimulated cardiac cells,* Science **214**, p.1350-1353.

HAAG, J. (1943). *Etude asymptotique d'oscillation de relaxation,* Ann. Sci. Ec. Norm. Sup. **60**, p.35-111.

HAAG, J. (1944). *Examples concrets d'étude asymptotique d'oscillation de relaxation,* Ann. Sci. Ec. Norm. Sup. **61**, p.73-117.

HABETS, P. (1978). *Relaxation oscillations in a forced Van der Pol oscillator,* Proc. Royal Soc. Edinburgh, sect. **A 82**, p.41-49.

HAGENDORN, P. (1981). *Non-linear Oscillations,* Clarendon Press, Oxford.

HAHN, H.S., A. NITZAN, P. ORTOLEVA and J. ROSS (1974). *Threshold excitations, relaxation oscillations and effect of noise in an enzyme reaction,* Proc. Nat. Acad. Sci. USA **71**, p.4067-4071.

HANSON, F.E. (1978). *Comparative studies of fire fly pacemakers,* Fed. Proc. **37**, p.2158-2164.

HAYASHI, C. (1964). *Non-linear Oscillations in Physical Systems,* McGraw-Hill, New York.

HÉNON, M. (1967). *A two-dimensional mapping with a strange attractor,* Comm. Math. Phys. **50**, p.69-77.

HERZEL, H., and W. EBELING (1985). *Chaos and noise in chemical models,* preprint Humbolt Univ. Berlin.

HILL, A.V. (1933). *Wave transmission as the basis of nerve activity,* Cold Spring Harbour Symp. on Quant. Biol. **1**, p.146-151.

HOLDEN, A.V. (1976). *The response of excitable membrane models to a cyclic input,* Biol. Cybern. **21**, p.1-7.

HONERKAMP. J. (1983). *The heart as a system of coupled nonlinear oscillators,* J. Math. Biol. **18**, p.69-88.

HONERKAMP. J., G. MUTSCHER and R. SEITZ (1985). *Coupling of a slow and a fast oscillator can generate bursting,* Bull. Math. Biol. **47**, p.1-21.

HOPPENSTEADT, F.C., and J.P. KEENER, *Phase locking of biological clocks,* J. Math. Biol. **15**, p.339-350.

IOOSS, G. and D.D. JOSEPH (1980). *Elementary Stability and Bifurcation Theory,* Springer-Verlag, New York.

ITO, K.,Y. OONO, H. YAMAZAKI and K. HIRAKAWA (1980). *Chaotic behaviour in great earthquakes: coupled relaxation oscillator model, billiard model and electronic circuit model of great earthquakes,* J. Phys. Soc. Japan **49**, p. 43-52.

JANSEN, M.J.W. (1978). *Synchronization of weakly coupled relaxation oscillators,* Math. Centre, Amsterdam, Report TW 180.

JOHANNESMA, P.I.M. (1984). Personal communication, Lab. Med. Phys. and Biophys., Cath. Univ. Nijmegen.

KAPLAN, B.Z., and I. YAFLE (1980). *An improved Van der Pol equation and some of its possible applications,* Int. J. Electron. **41**, p.189-198.

KAPLUN, S. (1957). *Low Reynolds number flow past a circular cylinder,* J. Math. Mech. **6**, p.595-603.

KARLIN, S., and H.M. TAYLOR (1975). *A First Course in Stochastic Processes,* Acad. Press, New York.

KEENER, J.P. (1981). *On cardiac arrythmias: AV -conduction block,* J. Math. Biol. **12**, p.215-225.

KEVORKIAN, J., and J.D. COLE (1981). *Perturbation Methods in Applied Mathematics,* Applied Math. Sci **34**, Springer-Verlag, New York.

KEITH, W.L., and R.H. RAND (1984). *1:1 and 2:1 phase entrainment in a system of two couped limit cycle oscilators,* J. Math. Biol. **20**, p.133-152.

KOPELL, N., and G.B.E. ERMENTROUT (1985). *Subcellular oscillations and bursting,* Preprint, Northeastern Univ.

KREIFELDT, J. (1970) *Ensemble entrainment of self-sustraining oscillators: a possible application to neural signals,* Math. Biosci **8**, p. 425-436.

KROGDAHL, W.S. (1960). *Numerical solutions of the Van der Pol equation,* **11**, p.59-63.

KÜNE, R.D. (1984). *Macroscopic freeway model for dense traffic: stop-start waves and incident detection,* in Ninth Int. Symp. on Transp. and Traffic Theory, VNU Science Press, p.21-42.

KURAMOTO, Y. (1975). *Self entrainment of population of coupled non-linear oscillators,* in Int. Symp. on Math. Problems in Theor. Phys., H. Araki,(ed.),Lecture Notes in Phys. **39**, Springer-Verlag, Berlin, p.420-422.

KURAMOTO, Y. (1984). *Chemical oscillations, waves, and turbulence,* Springer-Verlag, Berlin.

LASALLE, J. (1949). *Relaxation oscillations,* Quart. Appl. Math. **7**, p.1-19.

LAUWERIER, H.A. (1975). *A limit case of a Volterra-Lotka system,* Math. Centre, Amsterdam, Report TN 79.

LAUWERIER, H.A. (1985). Personal communication.

LE CORBEILLER, Ph. (1931). *Les systèmes autoentretenus et les oscillations de relaxation,* Conf. d'Actual, Scient. Industr. **27**, p.1-46.

LEVI, M. (1980). *Periodically forced relaxation oscillations,* in Global Theory of Dynamical Systems, Z. Nitecki and C. Robinson (eds.), Springer Lecture Notes in Math. **819**, p.300-317.

LEVI, M. (1981). *Qualitative analysis of the periodically forced relaxation oscillations,* Mem. Amer. Math. Soc. **244**.

LEVINSON, N., and O.K. SMITH (1942). *A general equation for relaxation oscillations,* Duke Math. J. **9**, p.382-403.

LEVINSON, N. (1949). *A second order differential equation with singular solutions,* Ann. of Math. **50**, p.127-153.

LIENARD, A. (1928). *Etude des ocillations entretenues,* Revue Générale de l'Electricité, **23**, p.901-946.

LINKENS, D.A. (1979). *Modelling of gastro -intestinal electrical rhythms,* in Biological Systems, Modelling and Control, D.A. Linkens (ed.), The Inst. of Electr. Engin., London, p.202-241.

LINKENS, D.A., and R.I. KITNEY (1982). *Mode analysis of physiological oscillators intercoupled via pure time delays,* Bull. Math. Biol. **44**, p.57-74.

LITTLEWOOD J.E. (1957a). *On non-linear differential equations of the second order:* III $y''-k(1-y^2)y'+y = b\mu k \cos(\mu+\alpha)$ *for large k, and its generalisations,* Acta Math. **97**, p.267-308.

LITTLEWOOD, J.E. (1957b). *On non-linear differential equations of the second order IV,* Acta Math. **98**, p.1-110.

LITTLEWOOD, J.E. (1960). *On the number of stable periods of a differential equation of the Van der Pol type,* IEEE **CT-7**, p.535-542.

LLOYD, N.G. (1972). *On the non-autonomous Van der Pol equation with large parameter.* Proc. Camb. Phil. Soc. **72**, p.213-227.

LORENZ, E.N. (1963). *Deterministic nonperiodic flow,* J. Atmos. Sci **20**, p.130-141.

LOPES DA SILVA, F.H., A. VAN ROTTERDAM, P. BARTS, E. VAN HEUSDEN and W. BURR (1976). *Models of neuronal populations: the basic mechanisms of rhytmicity,* in Progress in Brain Research, D. Swaab and M.E. Corner (eds.), Vol **45**, Biomed. Press, Amsterdam, p.281-308.

LOTKA. A.J. (1925). *Elements of Physical Biology,* Williams and Wilkins Comp., Baltimore.

LOZI, R. (1982). *Sur un modèle mathematique de suite de bifurcations de motifs dans la réaction de Belousov-Zhabotinsky,* C.R. Acad. Sc. Paris, Serie 1, **294**, p.21-26.

LOZI, R. (1983). *Modèles mathematiques qualitatifs simples et consistants pour l'étude de quelques systemes dynamiques experimentaux,* Thesis, Univ. of Nice.

MACGILLIVRAY, A.D. (1983). *On the leading term of the inner asymptotic expansion of Van der Pol's equation,* SIAM J. Appl. Math. **43**, p.594-612.

MAESS, G. (1965). *Quantitative Verfahren zur Bestimmung periodischer Lösungen autonomer nicht linearer Differentialgleichungen,* Abh. Deutsch Akad. Wiss., Berlin, klasse Math., Phys., Techn., Heft 3, p.33-40.

MARTZ, H.F., and R.A. WALLER (1982). *Bayesian Reliability Analysis,* Wiley, New York.

MATISOO, J. (1980). *The superconducting computer,* Sci. Am. May 1980, p.38-53.

MATSUMOTO, K. (1984). *Noise-induced order II,* J. Stat. Phys. **34**, p.111-127.

MATSUMOTO, K., and I. TSUDA (1983). *Noise-induced order,* J. Stat. Phys. **31**, p.87-106.

MAY, R.M. (1976). *Simple mathematical models with very complicated dynamics,* Nature **216**, p.459-467.

MAYERI, E. (1973). *A relaxation oscillator description of the burst generating mechanism of the cardiac ganglion of the lobster,* J. Gen. Physiol. **62**, p.473-488.

MINORSKY, N. (1962). *Nonlinear oscillations,* Van Nostrand, Princeton, New Jersey.

MISHCHENKO, E.F. (1961). *Asymptotic calculation of periodic solutions of systems of differential equations containing small parameters in the derivatives,* Am. Math. Soc. Transl. Series 2 **18**, p.199-230.

MISHCHENKO, E.F., and L.S. PONTRYAGIN (1960). *Differential equations with a small parameter attached to the higher derivatives and some problems in the theory of oscillation,* IRE Trans. on Circuit Theory, **CT-7,** p.527-535.

MISHCHENKO, E.F., and N. KH. ROSOV. (1980). *Differential equations with small parameters and relaxation oscillations,* Plenum Press, New York.

MITSUI, T. (1977). *Investigation of numerical solutions of some nonlinear quasiperiodic differential equations,* Publ. RIMS, Kyoto Univ. **13**, p.793-820.

MOSER, J. (1973). *Stable and Random Motions in Dynamical Systems,* Annals Math. Studies **77**, Princeton Univ. Press.

NAYFEH, A.H., and D.T. MOOK (1979). *Non-linear Oscillations,* Wiley-Interscience, New York.

NAYFEH, A.H. (1981). *Introduction to Perturbation Techniques,* Wiley, New York.

NEU, J.C. (1980). *Large populations of coupled chemical oscillators,* SIAM J. Appl. Math. **38**, p.305-316.

NEUMANN, D.A., and L.D. SABBAGH (1978). *Periodic solutions of Liénard systems,* J. Math. Anal. Appl. **62**, p.148-156.

NEWHOUSE, S., and J. PALIS (1976). *Cycles and bifurcation theory,* Astérisque **31**, p.43-140.

NICOLIS , G., and I. PRIGOGINE (1977). *Self-Organization in Nonequilibrium Systems,* Wiley, New York.

NIPP, K. (1980). *An Algorithmic Approach to Singular Perturbation Problems in Ordinary Differential Equations with an Application to the Belousov-Zhabotinskii Reaction,* Thesis, Diss ETH 6643, Zürich.

NIPP, K. (1983). *An extension of Tikhonov's theorem in singular perturbations for the planar case,* J. of Appl. Math. and Phys. (ZAMP) **34**, p.277-290.

NORTH, G.R. (1985). *The climate as natural oscillator,* Nature **316**, p.218.

Oka, H., and H. KOKUBU (1985a). *Constrained Lorenz-like attractors,* Japan J. of Appl. Math. **2**, p.495-500.

OKA, H., and H. KOKUBU (1985b). *Normal forms of constrained equations and their applications to strange attractors,* Proc. 24th IEEE conf. on Decisison and Control, Ft. Lauerdale Fl. , p. 461-466.

O'MALLEY, R.E. (1974). *Introduction to Singular Perturbation Theory,* Acad. Press, New York.

O'MALLEY, R.E. (1982). *Book review* of Kevorkian and Cole (1981) and Nayfeh (1981), Bull. Am. Math. Soc., New Series 7, p. 414-420.

OTHMER, H.G. (1985), *Synchronization, phase-locking and other phenomena in coupled cells,* in RENSING and JAEGER, eds. (1985). p.130-143.

PONTRYAGIN, L.S. (1961). *Asymptotic behaviour of the solutions of systems of differential equations with a small parameter in the higher derivatives,* Am. Math. Soc. Transl. Series 2 **18**, p.295-319.

PONZO, P.J., and N. WAX (1965a). *On certain relaxation oscillations: Confining regions,* Appl. Math. **23**, p.215-234.

PONZO, P.J., and N. WAX (1965b). *On certain relaxation oscillations: Asymptotic solutions,* SIAM J. Appl. Math. **13** , p.740-766.

PONZO, P.J., and N. WAX (1965c). *On the periodic solution of the Van der Pol equation,* IEEE Trans. on Circuit Theory, **CT-12,** p.135-136.

RAYLEIGH, Lord (1883). *On maintained vibrations,* Phil. Mag. **15**, p.229.

RENSING, L., and N.I. JAEGER, eds. (1985). *Temporal Order,* Springer-Verlag, Berlin.

RÖSSLER, O.E. , and E.WEGMANN (1978). *Chaos in Zhabotinskii reaction,* Nature **271**, p.89.

ROTHE, F. (1985). *The periods of the Volterra-Lotka system,* J. Reine und Angew. Math. **355**, p.129-138.

RUELLE, D., and F. TAKENS (1971). *On the nature of turbulence,* Comm. Math. Phys. **20**, p.167-192.

SANDERS, J. (1983). *The driven Josephson equation: an excercise in asymptotics,* in Asymptotic Analysis II, F. Verhulst (ed.), Springer Lecture Notes in Math. **985**, p.288-318.

SCHLUP, W.A. (1974). *I-V characteristics and stationary dynamics of a Josephson junction including the interference term in the current phase relation,* J. Phys. C: Solid State Phys. **7**, p.736-748.

SCHLUP, W.A. (1979). *Relaxation oscillations of a Josephson contact,* ZAMP **30**, p.851-857.

SCHLUP, W.A. (1981). *Relaxation oscillations in a cylindrical phase-space for a damped Sine-Gordon equation exhibiting turning points,* ZAMM **61**, p.4258-4258.

SEGEL, L.A. (1980). *Analysis of population chemotaxis,* in Mathematical Models in Molecular and Cellular Biology, L.A. Segel (ed.), Cambridge Univ. Press, Cambridge.

SHINOHARA, Y., A. KOHDA and T. MITSUI (1984). *On quasiperiodic solutions to Van der Pol equation,* J. Math. Tokushima Univ. **18**, p.1-9.

SIŜKA J., and I. DVOŔRAK (1984). *A generalization of Tikhonov theorem,* to appear in Casopis pro Pestováni Matématiky.

SMALE, S. (1967). *Differentiable dynamical systems,* Bull. Am. Math. Soc. **73**, p.747-817.

STANSHINE, J.A., and L.N. HOWARD (1976). *Asymptotic solutions of the Field-Noyes model for the Belousov reaction,* Stud. Appl. Math. **55**, p. 129-166.

STOKER, J.J. (1950). *Nonlinear Vibrations in Mechanical and Electrical Systems,* Interscience, New York.

STOKER, J.J. (1980). *Periodic forced vibrations of systems of relaxation oscillators,* Comm. Pure and Appl. Math. **33**, p. 215-240.

STORTI, D.W., and R.H. RAND (1986). *Dynamics of two strongly coupled relaxation oscillators,* SIAM J. Appl. Math. **46**, p. 56-67.

STRASBERG, M. (1973). *Recherche de solutions periodiques d'équations differentielles nonlinéaires par des méthodes de discretisation,* in Equations Differentielles et Fonctionelles Non-linéaires, P. Janssens, J. Mawhin and N. Rouche (eds.), Hermann, Paris, p.291-321.

STRITTMATTER, W., and J. HONERKAMP (1984). *Fibrillation of cardiac region and the tachycardia mode of a two oscillator system,* J. Math. Biol. **20**, p.171-184.

TAKENS, F. (1986). *Transitions from periodic to strange attractors in constrained equations,* Report ZW-8601, State Univ. Groningen.

THOM, R. (1972). *Stabilité Structurelle et Morphogénèse,* W.A. Benjamin Inc., Reading, Mass.

TIKHONOV, A. (1948). *On the dependence of solutions of differential equations on a small parameter,* Math. Sbornik **22**, p.193-204.

TOMITA, K. (1986). *Periodically forced nonlinear oscillators,* in Chaos A.V. Holden (ed.), Manchester Univ. Press, p. 211-236.

TORRE, V. (1975). *Synchronization on nonlinear biochemical oscillators coupled by diffusion,* Biol. Cybern. **17**, p.137-144.

TYSON, J.J. (1976). *The Belousov-Zhabotinskii Reaction,* Springer Lecture Notes in Biomath. **10** .

URABE, M. (1957). *Numerical determination of periodic solution of nonlinear system,* J. Sci. Hiroshima Univ., **A20**, p.125-148.

URABE, M. (1958). *Periodic solution of Van der Pol's equation with damping coefficient* $\lambda = 0(0.2)1.0$, J. Sci. Hiroshima Univ., **A21**, p.193-207.

URABE, M. (1960a). *Periodic solutions of the Van der Pol's equation with damping coefficient* $\lambda = 0 \approx 10$, IRE Trans. Circuit Theory 7, p. 382-386.

URABE, M. (1960b). *Remarks on periodic solutions of Van der Pol's equation.* J. Sci. Hiroschima Univ. **A24**, p.197-199.

URABE, M. (1963). *Numerical study of periodic solutions of the Van der*

Pol equation, in Int. Symp. on Nonlinear Differential Equations, J.P. LASALLE and S. LEFSCHETZ (eds.), Acad. Press, New York, p.184-192.

URABE, M., H. YANAGIWARA, and Y. SHINOHARA (1960). *Periodic solution of Van der Pol's equation with damping coefficient* $\lambda = 2 \approx 10$, J. Sci. Hiroshima Univ., **A23**, p.325-366.

VAN der POL, B. (1926). *On relaxation oscillations,* Phil. Mag. **2**, p.978-992.

VAN DER POL, B. (1940). *Biological rythms considered as relaxation oscillations,* Acta Med. Scand. Suppl. **108**, p.76-87.

VAN der POL, B. (1946). *Music and elementary theory of number,* Music Rev. **7**, p.1-25

VAN der POL, B., and J. van der MARK (1927). *Frequency demultiplication,* Nature. **120**, p. 63-364.

VAN der POL, B., and J. van der MARK (1929). *The heartbeat considered as a relaxation oscillation and an electrical model of the heart,* Arch. Neerl. Physiol. **14**, p. 418-443.

VATTA, F. (1979). *On the stick-slip phenomenon,* Mech. Res. Commun. **6**, p.203-208.

VELING, E.J.M. (1973). *An asymptotic approximation for the period of a Volterra-Lotka system,* Math. Centre, Amsterdam, Report TN 75 (Dutch).

VELING, E.J.M. (1983). Propositions to thesis "Transport by Diffusion", State Univ. Leyden.

VOLTERRA, V. (1931). *Lecons sur le Theorie Mathématique de la Lutte pour la Vie,* Gauthiers-Villars, Paris.

WALDVOGEL, J. (1983). *The period in the Volterra-Lotka predator-prey model,* SIAM J. Num. Anal. **20**, p.1264-1272.

WALDVOGEL, J. (1984). *The period in the Volterra-Lotka system is monotonic,* to appear in J. Math. Anal. Appl.

WALKER, J. (1983). *Thermal oscillators: systems that see-saw, buzz or howl under the influence of heat,* Sci. Am., Jan. 1983, p.116-121.

WEVER, R. (1965). *Pendulum versus relaxation oscillation,* in Circadian Clocks, J. Aschoff (ed.), North Holland, Amsterdam, p.74-83.

WEVER, R.A. (1979). *The Circadian System of Man,* Springer-Verlag, New York.

WHITHAM, G.B. (1974). *Linear and Nonlinear Waves,* Wiley, New York.

WHITTAKER, E.T., and G.N. WATSON (1927). *A Course of Modern Analysis,* Cambridge Univ. Press.

WIENER, N. (1958). *Nonlinear Problems in Random Theory,* MIT

Press, Cambridge, Mass.

WINFREE, A.T. (1967). *Biological rhythms and the behaviour of populations of coupled oscillators,* J. Theor. Biol. **16**, p.15-42.

WINFREE, A.T. (1980). *The Geometry of Biological Time,* Biomath. **8**, Springer-Verlag, New York.

WOLF, A., J. B. SWIFT, H.L. SWINNEY and J.A. VASTANO (1985). *Determining Lyanpunov exponents from a time series,* Physica **16D**, p. 285-317.

YANAGIWARA, H. (1960). *A periodic solution of Van der Pol's equation with a damping coefficient $\lambda=20$,* Fac. Sci. Hiroshima Univ. **A24**, p.201-217.

YOSHIZAWA, S., H. OSADA and J. NAGUMO (1982). *Pulse sequences generated by a degenerate analog neuron model,* Biol. Cybernet. **45**, p.23-34.

YPEY, D.L., W.P.M. van MEERDIJK, E. INCE and G. GROOS (1980). *Mutual entrainment of two pacemaker cells. A study with an electronic parallel conductance model,* J. Theor. Biol. **86**, p.731-755.

ZAROV, M., E.F. MISHCHENKO and N.H. ROZOV (1981). *On some special functions and constants arising in the theory of relaxation oscillations,* Soviet Math. Dokl. **24**, p. 672-675.

ZEEMAN, E.C. (1977). *Catastrophe Theory,* Selected Papers, 1972-1977, Addison-Wesley, Reading, Mass.

ZONNEVELD, J.A. (1966). *Periodic solutions of the Van der Pol equation* , Nederl. Akad. Wetensch. Proc. Ser. A **69**, p.620-622.

AUTHOR INDEX

SUBJECT INDEX

Applied Mathematical Sciences

cont. from page ii

39. Piccini/Stampacchia/Vidossich: **Ordinary Differential Equations in $\mathbf{R}^n$.**
40. Naylor/Sell: **Linear Operator Theory in Engineering and Science.**
41. Sparrow: **The Lorenz Equations: Bifurcations, Chaos, and Strange Attractors.**
42. Guckenheimer/Holmes: **Nonlinear Oscillations, Dynamical Systems and Bifurcations of Vector Fields.**
43. Ockendon/Tayler: **Inviscid Fluid Flows.**
44. Pazy: **Semigroups of Linear Operators and Applications to Partial Differential Equations.**
45. Glashoff/Gustafson: **Linear Optimization and Approximation: An Introduction to the Theoretical Analysis and Numerical Treatment of Semi-Infinite Programs.**
46. Wilcox: **Scattering Theory for Diffraction Gratings.**
47. Hale et al.: **An Introduction to Infinite Dimensional Dynamical Systems—Geometric Theory.**
48. Murray: **Asymptotic Analysis.**
49. Ladyzhenskaya: **The Boundary-Value Problems of Mathematical Physics.**
50. Wilcox: **Sound Propagation in Stratified Fluids.**
51. Golubitsky/Schaeffer: **Bifurcation and Groups in Bifurcation Theory, Vol. I.**
52. Chipot: **Variational Inequalities and Flow in Porous Media.**
53. Majda: **Compressible Fluid Flow and Systems of Conservation Laws in Several Space Variables.**
54. Wasow: **Linear Turning Point Theory.**
55. Yosida: **Operational Calculus: A Theory of Hyperfunctions.**
56. Chang/Howes: **Nonlinear Singular Perturbation Phenomena: Theory and Applications.**
57. Reinhardt: **Analysis of Approximation Methods for Differential and Integral Equations.**
58. Dwoyer/Hussaini/Voigt (eds.): **Theoretical Approaches to Turbulence.**
59. Sanders/Verhulst: **Averaging Methods in Nonlinear Dynamical Systems.**
60. Ghil/Childress: **Topics in Geophysical Fluid Dynamics: Atmospheric Dynamics, Dynamo Theory and Climate Dynamics.**
61. Sattinger/Weaver: **Lie Groups and Algebras with Applications to Physics, Geometry, and Mechanics.**
62. LaSalle: **The Stability and Control of Discrete Processes.**
63. Grasman: **Asymptotic Methods for Relaxation Oscillations and Applications.**